# Improvising Learning Imbalanced Data in Data Streams

By,

Bhowmick, Kiran

# ACKNOWLEDGEMENT

A blend of gratitude and pleasure that coats the joy and satisfaction which comes along with the successful completion of my work, would be blank unless I express my thankfulness to all those who have encouraged me and guided me as a beam of light at each and every step of this endeavour.

First and foremost, I would like to thank God almighty, who is always there for me and I strongly believe it only happens when He wants it to happen. I am writing this acknowledgement as He had wanted this thesis to be completed successfully.

I am thankful to Dr. Meera Narvekar, my guide, Professor and Head of Computer Engineering Department, D. J. Sanghvi College of Engineering, Mumbai. Her meticulous guidance, crucial advice, constructive criticism, constant encouragement and support have helped me sail through this journey of research. She has been always available and ready for discussion irrespective of the time and day. Her kind and gentle nature makes her easily approachable and cuts down the hesitation of asking or telling her something. She has always been prompt in reading the research papers inspite of her very busy schedule. My whole hearted gratitude goes to her for making this possible.

I would like to thank Dr. Hari Vasudevan, Principal, D. J. Sanghvi College of Engineering, Mumbai for his encouragement, guidance and valuable suggestions. His constant motivation helped me to complete this research as per the schedule. I would also like to thank Dr. A. C. Daptardar, Vice Principal (Admin) and Dr. Manali Godse, Vice Principal (Academic) for their motivation and support.

I take this opportunity to thank Dr. N. M. Shekokar for his valuable queries during the research discussions which helped a lot in the research. I am extremely thankful for his generosity and kind nature towards to me.

I am extremely grateful to the Annual Progress Review Committee members for their guidance and valuable inputs at every session. I am extremely grateful to Mr. Ramesh Sutar, Sr. Librarian for helping me with regular and timely plagiarism check for all the research publications and thesis.

To have cooperative batch mates who are constantly encouraging, suggesting, inspiring and even challenging you, is a blessing in disguise. I am fortunate to have such batch mates Ms. Aruna Gawade and Ms. Kriti Srivastava.

I take this opportunity to thank Ms. Chetashri Bhadane, my colleague and dear friend for her inspiration and suggestions during proof reading of the thesis. I am also extremely thankful to Dr. Abhijit Joshi, Professor, IT department for his constant motivation and suggestions.

I express my gratitude to my parents for their blessings, constant encouragement and motivation, my in-laws for their understanding and faith in me and my brother and sister-in-law who always wish for my well-being and success. This acknowledgement is incomplete if I do not express my gratitude to two most important persons in my life, my husband Pinaki and my dearest son Arthit. I thank Pinaki for being my pillar of strength, for giving unconditional love and support, for bearing with all my frustrations, anger, anxiety and joy during this journey. And my little angel Arthit, I thank him dearly for showing immense maturity and understanding, for constant encouraging and inspiring and for giving constant reminders to finish my thesis fast. I dedicate my thesis to Pinaki and Arthit.

I would also like to express my gratitude to all those people who have supported me directly or indirectly in completing this research successfully. Lastly, I thank everyone for being my companion in this tedious, exhausting but wonderful journey!

Kiran Bhowmick.

# TABLE OF CONTENTS

ACKNOWLEDGEMENT ................................................................................. i
ABSTRACT .................................................................................................. iii
TABLE OF CONTENTS ................................................................................. v
LIST OF TABLES ......................................................................................... ix
LIST OF FIGURES ....................................................................................... xi
LIST OF SYMBOLS AND ABBREVIATIONS ....................................... xiv

1. Introduction ............................................................................................. 1
   1.1 Relevance of the research problem ................................................. 3
   1.2 Applications of the research ............................................................ 4
   1.3 Need of the research ......................................................................... 5
   1.4 Overview of Learning in data streams ............................................ 5
      1.4.1 Traditional learning v/s Data streams learning technique ...... 6
      1.4.2 Incremental learning v/s Chunk-based learning ..................... 7
      1.4.3 Requirements of learning algorithms in data stream mining .... 9
   1.5 Imbalance data streams .................................................................. 10
      1.5.1 Challenges in Imbalance data streams ................................. 12
   1.6 Non-stationary data streams .......................................................... 13
      1.6.1 Types of Concept Drifts ....................................................... 15
      1.6.2 Challenges in Concept Drift ................................................ 16
   1.7 Research Problem .......................................................................... 17
   1.8 Research Process ........................................................................... 17
   1.9 Thesis Outline ................................................................................ 19
2. Review of Literature ............................................................................. 21
   2.1 Traditional approaches to deal with imbalance data .................... 21
      2.1.1 Data based approaches ......................................................... 22
      2.1.2 Algorithm based approaches ................................................ 23
      2.1.3 Hybrid approaches ................................................................ 24
   2.2 Literature review on imbalance data streams ............................... 24
   2.3 Analysis of Imbalance problem in data streams .......................... 28
   2.4 Gaps in Imbalanced data streams .................................................. 31

- 2.5 Concept Drift detection techniques..................................................................32
  - 2.5.1 Statistical based drift detection techniques....................................33
  - 2.5.2 Window based drift detection techniques......................................34
  - 2.5.3 Ensemble based drift detection techniques....................................35
- 2.6 Additional Literature review on concept drift ................................................36
- 2.7 Gaps in Concept Drift ......................................................................................38
- 2.8 Semi-Supervised Learning (SSL) approaches ................................................39
  - 2.8.1 Self Training ....................................................................................40
  - 2.8.2 Co-Training......................................................................................40
  - 2.8.3 Graph based training........................................................................40
  - 2.8.4 Generative Mixture Models..............................................................41
  - 2.8.5 Cluster and Label approach ............................................................41
- 2.9 Comparison of Semi-Supervised Learning approaches.................................41
- 2.10 Literature review on semi-supervised learning................................................45
- 2.11 Analysis of Semi-Supervised Learning ..........................................................48
- 2.12 Summary..........................................................................................................49

3. Research Problem, Objectives and Design ............................................................50
   - 3.1 Research Objectives.........................................................................................50
   - 3.2 Scope of the Research......................................................................................52
   - 3.3 Research Design...............................................................................................53
   - 3.4 Contributions of the research ..........................................................................54
   - 3.5 Summary..........................................................................................................55

4. Hybrid Ensemble Model for Imbalanced Data - HECMI ......................................56
   - 4.1 HECMI - The model ........................................................................................56
     - 4.1.1 Base classifier selection...................................................................58
     - 4.1.2 Dataset Partitioning ........................................................................58
     - 4.1.3 Ensemble of classifiers ....................................................................59
     - 4.1.4 Ensemble Result ..............................................................................59
   - 4.2 Datasets for HECMI ........................................................................................59
   - 4.3 Results and Conclusion for HECMI ................................................................60
   - 4.4 Limitations of HECMI.....................................................................................61
   - 4.5 Summary..........................................................................................................62

5. Concept Drift Detection and Adaptation model - CDACI......................................63
   - 5.1 CDACI – The Model .......................................................................................64

|  |  | 5.1.1 | Read the Input stream .................................................................... 64 |
|---|---|---|---|
|  |  | 5.1.2 | Class Imbalance detection ............................................................. 65 |
|  |  | 5.1.3 | Train a classifier .......................................................................... 65 |
|  |  | 5.1.4 | Classify new window ................................................................... 65 |
|  |  | 5.1.5 | Concept Drift Detection ................................................................ 65 |
|  |  | 5.1.6 | Concept Drift Adaptation ............................................................. 66 |
|  | 5.2 | Experimental setup for cdaci ................................................................... 66 |
|  |  | 5.2.1 | Datasets for CDACI ..................................................................... 66 |
|  |  | 5.2.2 | Experimentation for threshold and window size ........................... 67 |
|  | 5.3 | Results and Conclusion of cdaci .............................................................. 69 |
|  | 5.4 | Limitations of CDACI ............................................................................. 70 |
|  | 5.5 | Summary .................................................................................................. 71 |
| 6. | Semi-Supervised Clustering based Classifier Model - ASSCCMI ................... 72 |
|  | 6.1 | Need for Semi-Supervised learning ......................................................... 72 |
|  | 6.2 | Design decisions for semi-supervised learning model ............................. 73 |
|  |  | 6.2.1 | Chunk-based Ensemble approach for ASSCCMI ......................... 74 |
|  |  | 6.2.2 | Cluster and Label approach for ASSCCMI .................................. 74 |
|  | 6.3 | Top level Design of ASSCCMI ............................................................... 76 |
|  | 6.4 | Stream input module ................................................................................ 79 |
|  | 6.5 | Clustering based ensemble model ............................................................ 81 |
|  |  | 6.5.1 | SSL_EM clustering technique ...................................................... 82 |
|  |  | 6.5.2 | Macro cluster purity check ........................................................... 89 |
|  |  | 6.5.3 | Pure Micro cluster creation ........................................................... 91 |
|  |  | 6.5.4 | Label Propagation ......................................................................... 92 |
|  | 6.6 | Classification ........................................................................................... 95 |
|  | 6.7 | Concept Drift Detection and Adaptation .................................................. 96 |
|  |  | 6.7.1 | Drift Detection .............................................................................. 97 |
|  |  | 6.7.2 | Drift Adaptation .......................................................................... 101 |
|  | 6.8 | Refining the Ensemble ........................................................................... 101 |
|  | 6.9 | Summary ................................................................................................ 103 |
| 7. | Experimental Setup for ASSCCMI .................................................................. 104 |
|  | 7.1 | Experimental Setup ................................................................................ 104 |
|  | 7.2 | Dataset Description ................................................................................ 104 |
|  |  | 7.2.1 | Electricity dataset ........................................................................ 105 |

|  |  | 7.2.2 | Spambase dataset ........................................................................ 105 |
|---|---|---|---|
|  |  | 7.2.3 | Credit Card dataset ..................................................................... 106 |
|  |  | 7.2.4 | SEA dataset ................................................................................ 106 |
|  |  | 7.2.5 | Moving Hyperplanes dataset ...................................................... 107 |
|  |  | 7.2.6 | Summary of Datasets ................................................................. 108 |
|  | 7.3 | Evaluation Metrics ..................................................................................... 109 |
|  | 7.4 | Standard Algorithms for Performance Evaluation ..................................... 110 |
|  |  | 7.4.1 | Stream Algorithms for Comparison............................................ 111 |
|  |  | 7.4.2 | Semi-Supervised Learning Algorithm for Comparison ............... 113 |
|  | 7.5 | Parameter tuning ........................................................................................ 113 |
|  |  | 7.5.1 | Parameter Tuning for ASSCCMI ................................................ 114 |
|  |  | 7.5.2 | Parameter Tuning for Comparative algorithms .......................... 115 |
|  |  | 7.5.3 | Screenshots of Experimentation ................................................. 115 |
|  | 7.6 | Summary .................................................................................................... 121 |
| 8. | Research Findings and Analysis ............................................................................. 122 |
|  | 8.1 | Test Plan and Test Cases ............................................................................ 122 |
|  | 8.2 | Test Cases and Test Hypothesis ................................................................. 124 |
|  |  | 8.2.1 | Testing the Semi-Supervised Clustering approach SSL_EM ...... 125 |
|  |  | 8.2.2 | Test Case to assess the cluster purity check module of SSC ....... 135 |
|  |  | 8.2.3 | Test Cases for Drift Detection and Adaptation of ASSCCMI ..... 138 |
|  |  | 8.2.4 | Test Case for Overall Accuracy of ASSCCMI ............................ 150 |
|  |  | 8.2.5 | Test Case for Minority class accuracy of ASSCCMI .................. 160 |
|  |  | 8.2.6 | Test Cases for Execution Time of ASSCCMI ............................. 166 |
|  | 8.3 | Summary .................................................................................................... 168 |

Conclusion and Future Scope ................................................................................ 169
References ............................................................................................................... 173
List of Publications ................................................................................................. 183
Appendix ................................................................................................................. 185
Plagiarism report .................................................................................................... 192
Synopsis

# Chapter 1

Introduction

# 1. INTRODUCTION

Data today is generated at a mind boggling rate thanks to the technological advancements with cheaper hardware, accelerating growth in sensor technology, increasing usage of mobile devices and social media and increasing demand of staying connected. All of these are adding to the stock pile of data that is being gathered. This data is no longer static but is generated in a continuous manner having characteristics like it is infinite, massive, fast changing and temporal in nature. A large part of this data cannot be stored and must be processed as it arrives in real time [1]. Common examples of such a data that are seen in every day usage are web logs, computer network traffic, financial transactions, mail servers, social media platforms etc. Traditional data mining techniques are not sufficient to handle such a data. With increasing processing power and data storage capacity, new techniques for processing and extracting knowledge from this fast moving, volatile and voluminous data are needed [2]. The field of data stream mining provides efficient methods and techniques to address these problems.

Data streams are an unbounded sequence of data records, whose entry rate is usually high and whose dispersions frequently change. Algorithms used for analysing data streams must be able to process data in a very limited time frame and memory. This is due to the fact that, unlike traditional data, entire data streams cannot be stored in the memory [3]. Data is seen only once by the analysing algorithms when they process/analyze it and must be forgotten to make way for the incoming data. As a result efficient algorithms that can extract knowledge from the data streams in an online manner are the need of the hour.

Data streams that evolve over time suffer from the problem of concept drift wherein the underlying distribution of the data changes [4]. Concept drift generally occurs in non-stationary data streams. A model trained on a previous concept becomes irrelevant, when there is a change in the concept. As a result, the model either is scrapped or has to be retrained. Retraining a model can be informed or uninformed [5]. In uninformed drift detection, no specific drift detection takes place and the model retrains on misclassified instances. In informed drift detection, explicit drift detection mechanisms are employed and the model retrains on the instances/window

# Introduction

where drift is detected. The uninformed method leads to a lot of unnecessary retraining increasing the execution time. The informed method requires the drifts to be detected accurately. These methods have to deal with avoiding retraining when false alarms are raised. Improving the classifier accuracy in classification of non-stationary data streams is still the focus of many researchers.

In real streaming applications, a large volume of unlabeled data arrives at a high speed with only a fraction of labeled data. Labeling this unlabeled data is expensive as well as time consuming [6]. The stream classification methods operating on such data streams are therefore unsuitable in classifying the streams. Techniques that can exploit the advantages of the largely available unlabeled data are a topic of interest for current research.

Apart from the basic characteristics of data streams, they also tend to suffer from the problem of class imbalance. Class imbalance occurs when the number of instances of one class (majority / negative) is very high as compared to other class (minority / positive) [7]. The problem with classification of such datasets is that the classifier tends to misclassify the minority class instances and the cost of such misclassification is very high. In applications where the imbalance generally occurs, it is highly imperative to classify the minority class instances correctly [8]. The problem of class imbalance is important in many applications like fraud detection, spam filtering, life threatening disease prediction (cancer, tumour) etc., where it is extremely important to detect the fraudsters or email spams or positive cases for a disease before any major loss happens. Along with true negative cases it becomes absolutely necessary to predict the true positives as well.

This problem is well addressed for offline static data and many hybrid methods have evolved for improvising the accuracy of the model while classifying minority class instances. The problem of imbalance becomes severe in case of data streams. In real time applications such as medical diagnosis, credit card fraud detection and network intrusion detection the minority class are very rare and may not appear for a long time [9]. The classifier which has not seen the minority class might consider it as a noise and ignore or misclassify it. Classifying the minority class instances correctly is still a challenge in data streams primarily because the information about the classes and their distribution is not known a priori.

This chapter discusses about relevance of the research problem, different applications where the research problem is applicable, learning in data streams, requirements of learning algorithm in data streams, issues with imbalance data streams, issues with non-stationary data streams and need for the research along with a brief research problem definition.

## 1.1 RELEVANCE OF THE RESEARCH PROBLEM

This section discusses the importance/relevance of the research problem explaining where and how this particular research problem arises.

Consider a case of credit card transactions and the problem of detecting fraudulent and legal transactions for a bank. The transactions happening at every time period **t** can be considered as a data stream with each bank transaction as data instance. Now consider classifier learning from these transactions to classify it as a fraud or legal transaction. There are very few fraud transactions as compared to legal transactions. This is an example of imbalanced data streams. Many a times, not all the transactions will be labeled. The only available option is to do manual labeling or wait for the labels. Manually labeling of data is time consuming, cumbersome and usually expensive. Waiting for the labels to arrive at a later time is also not an option as in case of a fraudulent transaction the labels will be available after a period of 4-6 weeks when the customer complains about the fraud. This leaves the data stream with a small fraction of labeled data and a major chunk of data unlabeled. This scenario is the problem of classifying credit card transactions with imbalance and scarcely labeled data. Now, the fraudsters are always on the move for creating different types of frauds in credit card. These different types of frauds which are done over the time period are concept drifts. The classifier now has to adapt to correctly identify these new frauds thus leading to concept drift adaptation.

Another similar kind of application where labels are missing for most data is the spam detection classifier. Consider an email service user who receives a large number of emails every day which can be considered as stream of emails. Now, this user labels each of her emails as spam or ham, based on her interests. As a large number of emails are received every day the user takes the help of a classifier to detect spam and ham emails. Since the ham emails are more in number than the spam ones, this is an ideal case of imbalance data stream classification. Now, the users interests changes over time. For example, if she is buying a house then she will be interested in all the property sales related emails and will follow those making them hams. Once her house is bought, she is no longer interested in these property sales related mails and unsubscribes making them spam for her. This is a change in the description of class which leads to the problem of concept drift. Thus the underlying concept describing a class may change any time. The email system should be able to classify emails accurately considering the change in users description of spam and ham emails.

With technological advancements, the data is being generated continuously and at a faster rate than before. Like the credit card fraud and spam detection systems, there are many such real time applications which have to deal with data streams. These applications also face a similar problem of imbalance data. In the following section 1.2 a few of such applications are explained.

## 1.2 APPLICATIONS OF THE RESEARCH

There are many real world applications where the imbalanced data streams occur and the research problem is applicable. Some of the applications are listed and explained as follows:

**Network intrusion detection:** A network server handles a millions of requests in the form of data streams where each request is sent at a time $t_i$. Among these thousand requests placed every day to the network server some are legitimate and some are malicious. The number of malicious connections is very few and far less than the legitimate connections. This gives rise to the problem of imbalance. Detecting these malicious connections efficiently is the problem of network intrusion detection.

**Health care:** In health care or medical diagnosis problem, number of people infected with rare disease is insignificant as compared to healthy people, but the consequences of not detecting these patients are very severe. It is extremely vital to correctly detect and classify the rare diseases and the affected patients. Examples of such applications are cancer detection, tumour detection etc.

**Smaller scale computer networks:** These applications include network traffic, pattern of ATM transactions and customer purchase records for retail chains and online stores. The fraud detection in ATM transactions also suffers from the imbalance issue as the pattern of debit card fraudulent transactions are very few as compared to the legal transactions. Most of the internet traffic exhibits imbalance behaviour and are recently a focus of many researchers to provide a solution to the real-time imbalanced traffic classification. Customer purchase records are typical examples of data stream applications that suffer from concept drift.

**Oil spillage detection:** Detection and monitoring of oil spills in open sea and coastal waters is a major task which helps prevent pollution of sea beds and coastal fronts. Since the event of oil spilling is rare it is an imbalance problem. It is also a problem of data streams as the sea surface and coastal fronts are continuously monitored through radar images.

## 1.3 NEED OF THE RESEARCH

Though these are not new problems, learning algorithms are still struggling to achieve a good accuracy. Primarily because the type of the data, characteristics of the data, functioning of the learning algorithms, general assumptions about the system etc., all hinder the performance of the learners. Many researchers have provided solutions considering either imbalance or data streams. Very few have tried to give a solution for the combined problem of non-stationary data streams and imbalance both, but there is still scope for improvement. Providing a solution to such kind of problems is the need and motivation for this research.

## 1.4 OVERVIEW OF LEARNING IN DATA STREAMS

Learning in data streams is more challenging than in traditional datasets, primarily because of the nature of the data streams. Traditional datasets are static. So once collected they are stored and learning algorithms are trained on them by visiting and revisiting again and again. These can be massive in size, thus employing learning in distributed or parallel environment. Data streams however are fast moving, fast changing, potentially infinite and massive. The manner in which algorithms process data while learning from static data and data streams is quite different.

### 1.4.1 Traditional learning v/s Data streams learning technique

In traditional static learning, algorithms are first trained on a training set to create a model which is then validated on a testing set. The model is then used to predict/classify an unseen data as shown in Figure 1-1 below. The training set, test set and the model all reside in main memory.

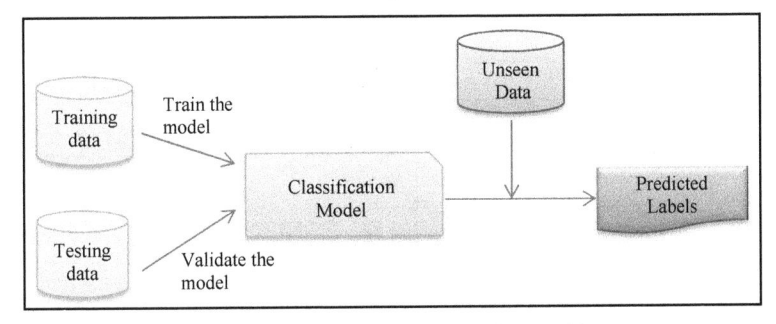

**Figure 1-1: Traditional Classification model**

In data streams, data is being read continuously from a source. A model is trained on the read data and is ready for prediction. As shown in Figure 1-2 below:

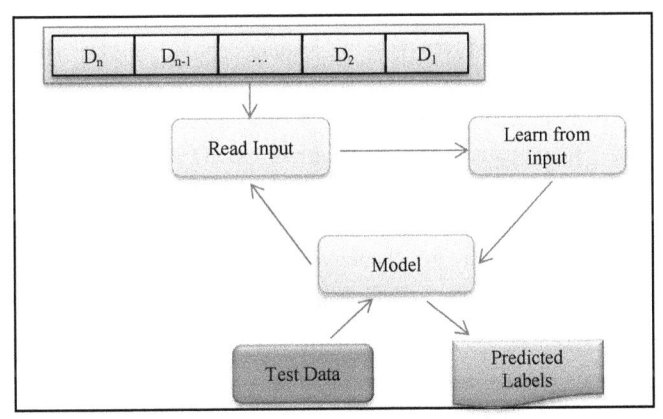

**Figure 1-2: Data streams Classification model**

When a new data comes the learning model is updated and prediction happens on this new model. At any given time **t**, prediction depends upon the model that has learnt from a data in (**t-1**).

A comparison of the characteristics of the mining model in traditional and stream mining is shown in the Table 1-1 below [10].

Table 1-1: Comparison of traditional and stream mining

| Algorithm characteristics | Traditional Mining | Stream Mining |
|---|---|---|
| No. of passes | Multiple | Single |
| Processing time | Unlimited | Restricted |
| Memory usage | Unlimited | Restricted |
| Type of result | Accurate | Approximate |
| Concept | Static | Static or Evolving |
| Distributed | No | Yes |

Due to the huge volume of data streams and its need for fast processing, these cannot be stored in the memory for further analyses and should be analyzed through methods such as incremental learning or chunk (batch) based learning approaches. The comparison between these two learning approaches is explained in the following section.

### 1.4.2 Incremental learning v/s Chunk-based learning

In the incremental learning process, the system learns from each single data sample as it arrives. Whereas, in a chunk-based learning process, system learns from a chunk of 'n' data samples. The performance of a chunk-based learning approach is similar to the incremental learning approach. One disadvantage of incremental learning is that it may require massive number of samples to learn, and may not adapt naturally to the changes in the data, whereas a chunk-based learning automatically deals with changes in the data stream. Incremental learning also known as online learning generally employs a single model / classifier that are updated incrementally.

# Introduction

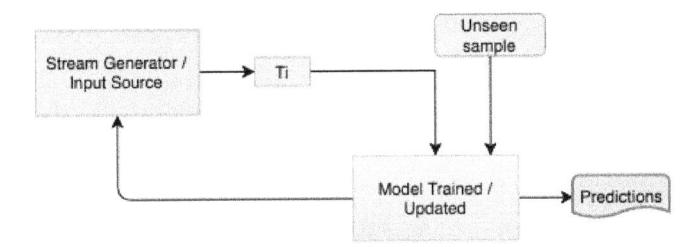

**Figure 1-3: Incremental Learning**

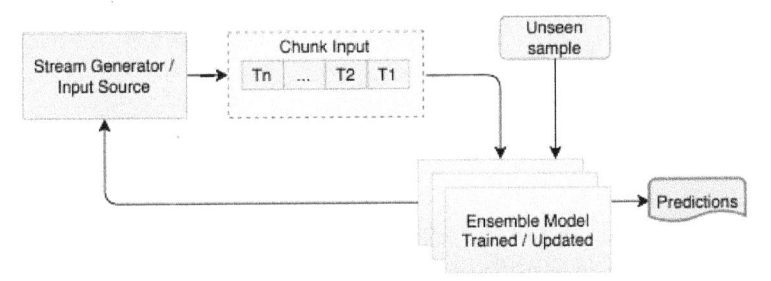

**Figure 1-4: Chunk-based learning**

Chunk-based learning also known as batch-based learning may employ single or ensemble of classifier. Ensemble model have proved to be advantageous even in conventional static learning. Instead of a single classifier an ensemble of classifiers is trained that predicts a class using some consensus like majority voting, weighted average etc. Ensemble results have proven to be more accurate than the single classifier [7]. Ensemble learning can also be used in data streams with chunk-based learning. Each model in the ensemble can train on smaller data-chunks independently making the cost of classification low with only limited memory usage at one time [11].

A comparison of chunk based learning and incremental learning is shown in Table 1-2 below:

**Table 1-2: Chunk-based Vs Online learning**

| Parameter | Chunk based Learning | Online / Incremental learning |
|---|---|---|
| Size of data | Deciding on size of chunk is difficult. | No batch size required. |
| Processing | Batch not processed until it is full. | Data is processed as it arrives. |

Introduction

|  | Slow as compared to Incremental. | Fast and efficient as compared to chunk based. |
|---|---|---|
|  | It deals with the changes in data stream automatically and it is able to detect and adapt to concept drift. | May require massive number of samples to learn, and may not adapt naturally to the changes in the data. |
| Variety in data | The batch might contain only one class of data so modelling is difficult. | At an instant, the stream might also contain data which it hasn't seen earlier. |
| Classification | Done after the batch is full. | Done instantly. |
| Training classifiers | Need to rebuild model from scratch using old and new instances. | Update model incrementally on current instances forgetting the previous ones. |

### 1.4.3 Requirements of learning algorithms in data stream mining

Any learning algorithm must meet certain requirements in order to be suitable for data stream mining. These requirements place constraints on the functioning of an algorithm in a data stream environment [12] [13] [14] [15]. The requirements are as follows:

1. **An instance / example is processed one at a time and accessed only once at most:** Data streams are received as a sequence of instances / examples one after another. These instances are to be processed as they are received. An algorithm might process or ignore the instance but in any case these instances will be discarded without the ability to retrieve it again. Algorithms may choose to store these instances internally for short term. Eventually the algorithm might have to forget the stored data. Algorithms that are not able to conform to the single pass rule i.e. accessing data only once are not applicable for data stream processing.

2. **Availability of only limited memory for processing:** Data streams are unbounded infinite sequence of data. The data available is generally much more than the available working memory and the processing of this large amount exhausts the limited memory. Learning algorithm should be able to process the data in the available capacity of working memory.

3. **Execution in limited amount of time:** The classification algorithms should be fast enough to be able to process data faster than or at least as fast as the arrival rate of the data. Failing to do so will result in the loss of the arriving data.

4. **Any time prediction:** The classification algorithms should be able to create best models from the data that it is seeing once and should be able to produce results efficiently. More emphasis is given on producing the results faster rather than accurate. An approximate result as opposed to accurate is acceptable in data streams mining.

Traditional classification algorithms that are used in mining of static datasets do not have to function in such stringent rules. These algorithms are provided with a dataset that stays in the memory as long as it is needed, thus allowing these algorithms to iterate over and over again till a perfect model is learned. Hence the accuracy of such algorithms is also expected to be the best. The learning requirements for data streams impose implementation challenges on the algorithms used in data stream mining.

## 1.5 IMBALANCE DATA STREAMS

The data streams in some of the applications mentioned in Section 1.2 suffer from the problem of skewed distributions where one class has more (majority / negative class) representation than another (minority / positive class). Imbalance is caused when one class data is very rare to find like the case of finding a cancerous patient as opposed to a healthy person. In such cases, the number of instances representing the majority class is far more than the minority class. Sometimes the ratio of majority to minority class can be as high as 1000:1. These are data with severe imbalance.

Shown in Figure 1-5 is a sample of Spambase dataset where the green dots represent the ham emails and the red dots represent spam emails. Here the number of spam emails is less than the non-spam or ham emails and the ratio of majority to minority class is 1.5:1.

Another example where there is an imbalance in the data is Credit Card transactions data. Shown in Figure 1-6 is a sample of first 1000 instances of Credit Card dataset where the green dots represent legal transaction and the red dots represent fraudulent transaction. Here too, the number of fraudulent transactions is far less than the legal transactions and the ratio of majority to minority class is 2.3:1.

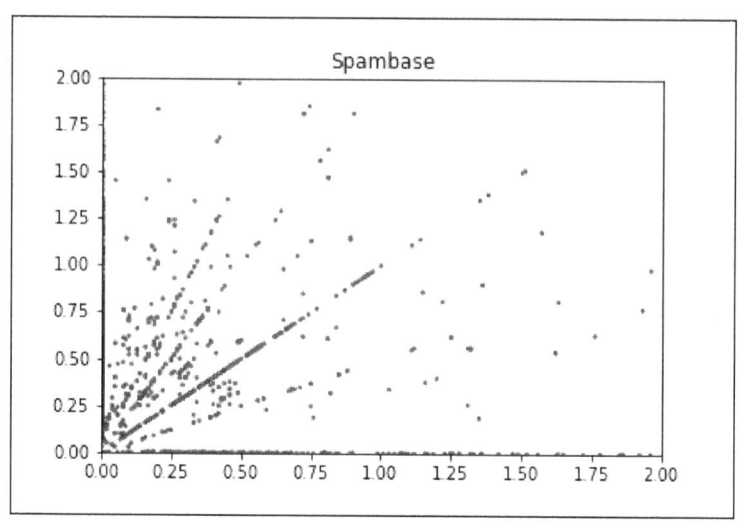

Figure 1-5: Data distribution of Spambase dataset

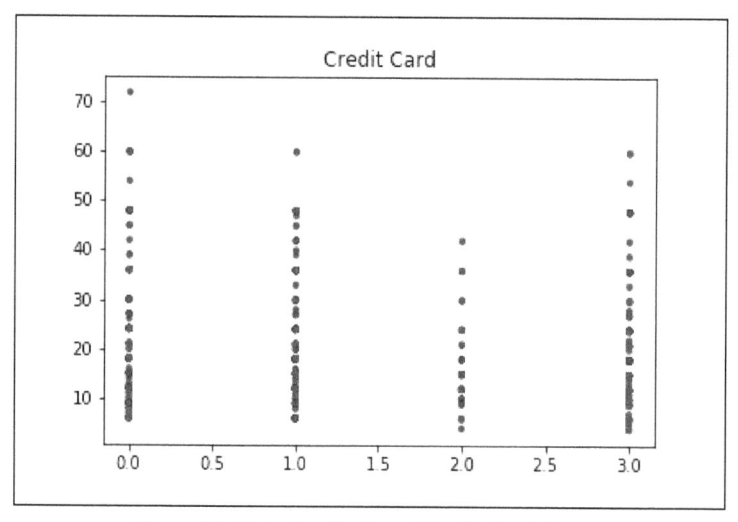

Figure 1-6: Data distribution of Credit Card dataset

When a classifier is trained on an imbalance data, it learns the majority class with good accuracy. However, due to scarcity of minority class data, the classifiers accuracy for minority class falls considerably. In many such cases, the classifiers accuracy can be as high as 97%. But this accuracy is only the measure of correctly classifying the true negatives i.e. majority class. It indicates that the classifier is able to classify the majority class instances 97% of the time. The remaining 3% mostly belongs to the misclassification of minority class instances. If the classifier is supposed to predict whether a patient is cancerous or not, then this misclassification will lead to serious consequences. A cancerous patient will not be declared as cancerous and will be devoid of life saving treatments.

The imbalance data in data streams also affects the classifier's performance like it does in static data. This happens because the classifier becomes more biased towards the majority class as the classifier has been trained on more number of these instances as against the very few instances of the minority class [16]. Any classifier trying to classify imbalanced data should try to improve the classification of the minority class.

### 1.5.1 Challenges in Imbalance data streams

Classifying imbalance data in data streams can be challenging due to various reasons [9] [17] [18] :

1. Data stream is not stored and has to be processed as soon as it arrives. This **requires prior information** about the description of majority and minority classes. Lack of such information and the uncertainty about the imbalance status makes the classification of data streams difficult.

2. In case of online/incremental learning, the **classifier won't see a minority class instance** for a long time while processing data in streams. As a result, training the classifier with minority class will still be a problem.

3. Classification of **minority class** is affected more by **noisy data** as compared to the majority class. This happens due to the fact that minority class instances are very few in number as compared to majority class and also the classifier do not see the minority class for long as a result it tends to treat the minority class as noise.

4. **Traditional approaches** of handling imbalance in static data have proven to be **insufficient in handling imbalance data streams**. Novel adaptive techniques that can determine imbalance dynamically have to be determined for data streams.

5. **Data streams suffer from drifts** wherein the underlying distribution of the concept changes. This change can happen in majority or minority class. Detecting this change in concept with imbalance data in data streams is a major challenge.

## 1.6 NON-STATIONARY DATA STREAMS

Data streams are also considered to be volatile making them dynamic in nature i.e. the environment is ever changing [1]. In such scenarios, even if the old data is stored temporarily it is of very limited use as the change in data has induced different types of changes viz., changes in the features of information, changes in the target and change in the underlying distribution. Concept drift occurs when there is a change in the underlying distribution of the data [19]. The change can be in the target variable or in the distribution of the input variable. The most common example of concept drift is the problem of spam filtering. The characteristics of spam emails are user specific and are evolving with time. As mentioned in the email example of section 1.1, after the user has bought a house she is no longer interested in the mails regarding property sales and all the property sales mail are marked as spam. So the characteristics of the ham emails have changed over time.

A classification decision in a classification problem is made according to the posterior probabilities of the classes, which is given by:

$$p(y|X) = \frac{p(y).p(X|y)}{P(X)} \qquad (1.1)$$

Where,

p(y) – prior probabilities of classes

p(X|y) – conditional probability for all classes y = 1 … c

$$p(X) = \sum_{y=1}^{c} p(y).p(X|y)$$

In a classification problem, the drifts can occur due to the following reasons [20] [21] :

1. Change in the prior probability of the class i.e. p(y)

2. Change in the class conditional probability p(X|y)

3. Change in the posteriori probability p(y|X)

From the changes listed above, only those that affect the prediction performance require adaptation. Based on the above reasons of drift occurrence, drifts can be distinguished as real drift and virtual drift:

1. **Real Concept Drift:** Real drifts refer to the changes in the posteriori probability p(y|X). These drifts affect the prediction performance and are required to be adapted.

2. **Virtual Concept Drift:** Virtual drifts refer to the changes in the incoming data i.e. p(X) which do not affect p(y|X). These drifts do not affect the prediction performance and are not required to be adapted.

Considering the example of spam filtering, the task is to classify the email as spam or ham. Now, when the user is looking for buying a house, all mails related to property sales are ham for her and mails related to holiday homes are spam. The mails can arrive from different property sellers so the style of email changes nevertheless it still is ham for her. This is virtual drift where change in the incoming data p(X) does not affect p(y|X).

Now, there can be more number of mails regarding property sales and less of holiday homes. This is an example where the prior probabilities p(y) have changed but the situation is still same.

When the user has bought a property and she is no longer interested in mails related to property sales but starts looking for holiday homes for vacation. Now here the property sale mails have become spam and holiday homes related mails are ham. This scenario is real concept drift. The mails are same but the concept for the user has changed.

It should be also noted that concept drifts are different than trends. Trends are predictable and also periodical e.g., fashion trends follow change in season, user

# Introduction

buying pattern are predictable and depend on various factors like available discounts, seasons, fashion trends etc. However, drifts are not periodical and predictable. The reasons of drifts are not known until they occur. Hence, most drift detection techniques are based on the classifiers performance. When the performance degrades it is assumed that the data on which the classifier was trained has changed i.e. drift has occurred. These drifts can be of various types as described below in Section 1.6.1.

## 1.6.1 Types of Concept Drifts

There are different types of drifts that can occur either individually or together in a data stream. The Figure 1-7 shows a representation of these different drifts. The circles are instances, the blue circles represent instance of one class and the red represents another.

1. **Sudden or abrupt drift:** When a drift happens instantly or suddenly changing the concept irreversibly, then it is called a sudden drift. E.g. season change in sales or replacing a sensor with new sensor of different calibration are all examples of sudden changes.

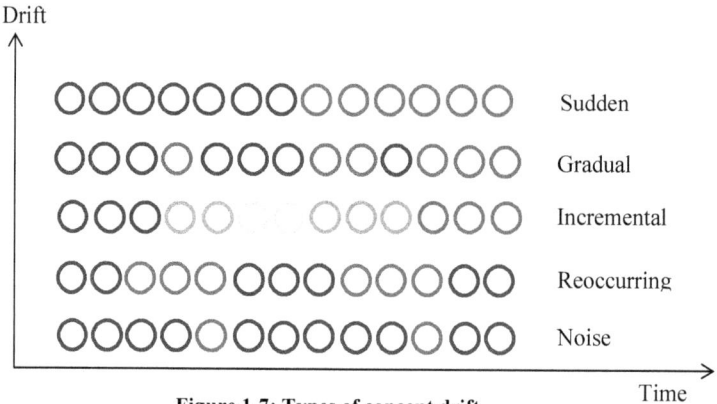

**Figure 1-7: Types of concept drift**

2. **Gradual drift:** When the change happens slowly over time, then it is a gradual drift. In gradual drift, the concept keeps going back to the earlier concept. For e.g., in spam filtering example the user's interest in property sales and holiday

homes keeps dwindling back and forth. Another example is user's interests in news keeps changing from sports to election.

3. **Incremental drift:** Similar to gradual drift, this drift also happens slowly over the time. In this, a concept changes from one class to another through many intermediate stages. For e.g., before a sensor totally goes bad it starts wearing off slowly. Another example is the price growth due to inflation.

4. **Recurring drift:** Changes that are temporary and keeps reverting back to earlier concepts. For e.g., previous trends in fashion keeps coming back after some time.

5. **Noise:** Noise is not related to change in concept but mere insignificant fluctuation and should be filtered out.

### 1.6.2 Challenges in Concept Drift

Learning algorithms in the non-stationary environment need to have mechanisms to detect and adapt to the changing / evolving environment. Like the requirements of any learning algorithms for data streams, these should also be able to:

1. **Adapt to evolving data** – the underlying distribution of the data changes dynamically over time. The change in the data generally happens unexpectedly and unpredictably. In most real cases these changes are not known in advance.

2. **Detect drift as soon as it occurs and adapt to it:** Detecting drifts can be uninformed wherein the learning algorithms subtly keep adapting to the drifts by learning the misclassified instances and modifying itself. Drift detection can be informed where the learning algorithms adapt to the drift only when it is detected. This prevents the algorithms from learning unnecessarily. Adaptation to the drift happens by either updating the existing model or forgetting the previously learned model to learn a new one.

3. **Perform using available memory and limited time:** Since the data is huge and cannot be stored, learning algorithms have a restriction on using memory resources and must operate in time less than the arrival rate.

4. **Distinguish between noise and outliers:** Noise and outliers are not change in the concept. Learning algorithms must be able to distinguish noise and outliers from concept drift and must be robust to noise.

## 1.7 RESEARCH PROBLEM

Classification in data streams has to deal with non-stationary data where in the distribution of the data changes considerably over-time and there is an unavailability of class labels apriori. The labeling of classes would impose tremendous cost for data streams. The cost in terms of time, effort and expenses required to manually label the unlabeled data is high. In case of imbalance data, the information about the ratio of imbalance is not known beforehand. This makes the imbalance non-stationary data classification a difficult task. The absence of a framework that handles all these issues gave a direction for a new research problem which is stated in a broader sense as below:

**"Learning to classify imbalance data streams with concept drift"**

The aim of the research is to design a framework to classify data streams that are non-stationary and suffer from the problem of imbalance. The framework should conform to the requirements of learning algorithms in streaming environment.

The research process followed in this research is explained below. The detailed problem statement and the research objectives are discussed after analysing the problem in detail and considering the gaps in the literature review.

## 1.8 RESEARCH PROCESS

The main objective of the research is to improvise the learning for data streams considering the class imbalance problem. This would be achieved by developing an adaptive learning model that is able to perform classification for imbalance data streams.

The type of research methodology used is an applied analytical and experimental as it intends to analyze the performance of the proposed model by evaluating it against different standard methods and experimenting it on different benchmark datasets. The aim is to identify a suitable technique to deal with the problem of imbalance data that affects many different applications discussed in section 1.2.

The research process followed is shown in Figure 1-8.

Introduction

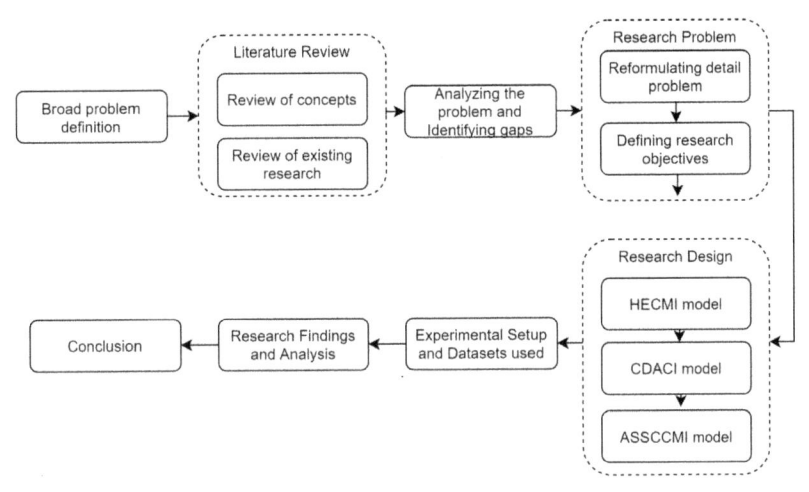

**Figure 1-8: Research Process**

After defining a broader problem definition, related concepts about imbalance data, concept drift and semi-supervised learning are reviewed to get a conceptual understanding about existing methods. Then a review of literature is done to find out about existing research and also analyze the problem of imbalance with respect to data streams.

Gaps are identified in the existing research with respect to the problem defined earlier. The problem is redefined with clearly stated research objectives. The research process begins by designing models for individual problems and then analysing benefits and limitations of each of these models design a framework that will be able to justify the research problem as a whole. A hybrid ensemble model for classifying imbalanced datasets HECMI is first designed. This model deals with imbalanced data using sampling for balancing the data and standard classification techniques to classify the data. The novelty is in the way the data is partitioned and selected for oversampling. This model improves the performance for imbalanced datasets as compared to the state of the art classification techniques.

A second model is designed to classify imbalanced data streams using incremental learning. The model CDACI is able to detect drift in the data and adapt to it. It uses sampling to balance the data that is read in a window of instances. It also uses standard classification techniques to perform classification. It is able to improve the classification by detecting and adapting to drifts as compared to the state of the art classification techniques.

These two models are then analyzed to identify the benefits and limitations in them. ASSCCMI an adaptable semi-supervised learning algorithm that is able to overcome the limitations of HECMI and CDACI and also improve the performance of classifier to a considerable extent is designed. The model uses semi-supervised clustering based classifier to perform classification on scarcely labeled imbalanced data streams and an adaptive window based statistical testing method to detect and adapt to concept drifts in it.

The model is then evaluated against 7 different benchmark datasets; all having different sizes, number of attributes, imbalance ratio and also types of concept drifts. HECMI and CDACI are evaluated on selective datasets suitable for their problem statement from these 7 datasets. The performance of HECMI and CDACI are compared against standard supervised classification techniques whereas ASSCCMI is compared with standard stream classification and an existing semi-supervised learning framework. Test cases and hypotheses are used for evaluating ASSCCMI and prove its improvement.

## 1.9 THESIS OUTLINE

The overall organization of the topics and chapter-wise overview of the content is explained below.

Chapter 2 discusses the existing approaches for dealing with imbalanced data, analysing the occurrence of imbalance in data streams, detail literature review of the existing research on imbalance in data streams and the gaps identified in the literature review. It also discusses the different approaches to deal with concept drift, literature review and the gaps identified. It also discusses different semi-supervised learning approaches discussing their pros and cons. The chapter also covers the literature review of related research done in the field of semi-supervised learning and data streams.

Chapter 3 redefines the problem definition, discusses the research objectives, and defines the scope of research and the research contributions.

In Chapter 4, a novel ensemble model, HECMI, to handle imbalanced data sets is explained with its working, experimentation, results and limitations.

In Chapter 5, working of a novel drift detection and adaptation model, CDACI, is explained with its experimentation, results and limitations.

Chapter 6 covers in detail the recommended Semi-Supervised learning based model, ASSCCMI, explaining in detail each and every module and the respective algorithms.

Chapter 7 describes the experimental setup done to evaluate the performance of the proposed research design ASSCCMI, detailed description of datasets used for evaluating the designs and evaluation metrics. The different stream classification methods used for comparing the performance of ASSCCMI along with the parameter tuning for each of these methods are discussed.

Chapter 8 discusses the research findings and their analysis using different test cases are also explained in this chapter.

Lastly, the thesis covers the conclusion and future scope of this research, followed by list of publications done, references cited and appendix.

# Chapter 2

Review of Literature

# 2. REVIEW OF LITERATURE

The review of literature helps in understanding the conceptual details of the research problem, current research done by fellow researchers in the domain, analysing the problem and identifying the gaps that needs to be addressed.

In this chapter, existing approaches, detailed review of literature and gap analysis of literature review to deal with respective problems of imbalance data streams and concept drift are discussed. Section 2.1 discusses about the approaches to deal with the imbalance issue, literature review about the research in the domain of imbalance problem is done in section 2.2, analysis of how the problem affects in data streams is done in section 2.3 and section 2.4 discusses the gaps identified in imbalance data streams. Section 2.5 discusses the different drift detection techniques, section 2.6 discusses the literature review on concept drift and section 2.7 discusses the gaps in the literature review of concept drift that needs to be addressed. This is followed by a review for Semi-Supervised Learning (SSL) techniques in Section 2.8, comparison of the SSL techniques in Section 2.9, review of existing literature in semi-supervised learning and data streams in Section 2.10 and analysis of SSL in Section 2.11

## 2.1 TRADITIONAL APPROACHES TO DEAL WITH IMBALANCE DATA

Any learning algorithm performs efficiently when the data it is being trained on is well balanced i.e. the ratio of number of instances belonging to both classes in case of binary classification is equal. If not so, then there are higher chances of the learning algorithm being biased towards the majority class [7].

Some traditional approaches to deal with imbalance are applied by many researchers. These approaches work very well when the data is static. These approaches can be applied to data streams but the inherent properties and requirements of data streams should be considered.

Imbalance is handled by using either data based approaches, algorithmic based approaches or both. Generally, either the data is balanced by using techniques like under sampling, over sampling and both or the imbalance is handled using algorithmic methods like cost-sensitive methods or ensemble methods [22]. The different approaches for handling imbalanced data are shown in Figure 2-1 below.

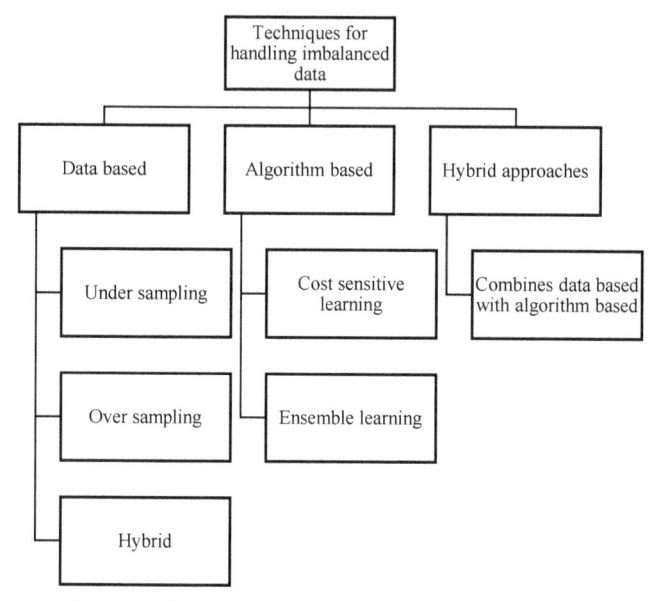

**Figure 2-1: Traditional approaches of dealing with imbalanced dataset**

The approaches to deal with imbalanced data are described below:

### 2.1.1 Data based approaches

The data based approaches include re-sampling or feature selection at the data level. In re-sampling, the data is balanced either by adding or removing samples. The techniques include oversampling where the instances of the minority class are added and under sampling where the instances from majority class are removed to balance the classes. A hybrid of both over sampling and under sampling can also applied jointly to balance the data. Under sampling causes classifiers to miss out on important concepts whereas oversampling simply replicates data [23]. Smart approaches like SMOTE proposed by Chawla et. al., in [24] creates synthetic examples using the minority instances to balance the training chunk.

Problem with oversampling is that since synthetic instances are added, it may add meaningless instances to the sample making learning algorithm learn from meaning less data. Whereas in undersampling, removal of instances may remove meaningful data making learning algorithm miss out on important concepts. Hybrid sampling approaches are also proposed that combine oversampling and under sampling to balance the data.

In feature selection based techniques, a subset of k-best features are selected by different method that optimizes the classifier performance. This technique reduces the risk of minority class being treated as noise. Feature selection is also used in dimensionality reduction [18].

## 2.1.2 Algorithm based approaches

The algorithm level approaches concentrate on modifying the existing learners and reduce their bias towards majority class [25]. The methods include cost-sensitive learning and ensemble learning at the algorithmic level.

Cost-sensitive learning technique takes into account the misclassification cost by assigning higher-cost to the misclassification of minority class. Considering a fixed cost model is one of the solutions, but in most cases the misclassification error costs is unknown. Also for many real time applications it is difficult to set the actual values for the cost matrix.

Ensemble based methods are another popular methods for class imbalanced techniques. Multiple classifiers are combined to provide accurate and robust predictions. The most popular ensemble techniques are bagging, boosting and random forest. Ensemble methods are able to deal with the phenomenon of concept drift in data streams [26] [27]. However, while considering ensemble learning the diversity of the classifiers is an important aspect. If all classifiers make same predictions, then the ensemble is no better than a single classifier. Selecting appropriate classifiers to form the ensemble is an important decision. Since the ensemble methods focus on the performance of the classifier and not on the distribution of the data, a combination of data-based methods like under or over sampling and ensemble methods are proposed in many of the literature.

The following Table 2-1 compares ensemble learning with cost-sensitive learning discussing their pros and cons [28].

**Table 2-1: Ensemble Vs Cost-sensitive learning**

| Parameter | Ensemble Learning | Cost-sensitive learning |
|---|---|---|
| Variety in data | Good for imbalanced data | Can be used for imbalance |
|  | Can be used for data streams | Better approach for data streams |
| Classification | Good classification results | Better classification results |
| Training classifiers | Each individual classifier performs classification on the unseen data. The final result is the combination of prediction of each model like weighted average or majority polling. | A cost matrix is created where the misclassified samples are given higher costs than the correctly classified ones to make adjustments in the classifier learning. |
|  | Flexible and adaptive | Cost matrix can be made adaptive to some extent |
|  | Adapts to concept drift | Doesn't adapt very well |
| Processing | Diversity of the classifiers in the ensemble is major issue. | Assigning the misclassification cost is difficult and usually depends on the type of data. |

Though compared to the re-sampling methods cost sensitive techniques are computationally more efficient for data streams [4] but their inefficiency to adapt to concept drift suggests ensemble methods for problems dealing with imbalance non-stationary data streams.

### 2.1.3 Hybrid approaches

The hybrid approaches include a combination of data based as well as algorithm based approaches. In these approaches, the data is balanced using one of the sampling methods viz., either over-sampling or under-sampling or both. Then on the balanced data an algorithmic based model is trained and used for classification. Many of the models suggested in the literature use the hybrid method. It is seen that the performance of the model considerably increases with the application of hybrid approach.

## 2.2 LITERATURE REVIEW ON IMBALANCE DATA STREAMS

The literature review for imbalance data streams is separated into related research papers by various authors and discussion that focus on issues and challenges in imbalance, handling imbalance at data level, handling imbalance at algorithmic level followed by existing frameworks that have tried to solve the problem of imbalance in data streams.

In [16], Ryan et. al., have discussed about the challenges of learning from non-stationary environment together with the imbalanced data. They have also suggested about the need of a framework that will deal with non-stationary imbalanced data. The need to design one unified classifier that can deal with different types of drifts also has been highlighted by the authors. Use of ensemble learners to deal with class imbalance problems is also suggested by the authors

The authors in [17] discuss issues and challenges in static as well as streaming data. The paper also discusses about various methods of dealing with issues in imbalance like multiclass, non-availability of data labels, fading class issues etc. The paper also discusses about various applications and domains where imbalance data causes severe problems. These include the distributed environment provided by Hadoop and Spark. Use of cost sensitive online learning is more efficient for big data streams as suggested in the paper.

In [29], provides a detail survey of research on ensemble classification of data streams. The paper provides a detail review of various problems in data streams that is class imbalance, concept drift, diversity of classifiers in ensemble and need for evaluation measures. The authors have also hinted over the possibility of using semi-supervised learning for exploiting the advantages of unlabeled data.

Earlier works in **imbalanced data focused on balancing the datasets** by using various sampling techniques for balancing the data. SMOTE [24] is the first of the smart sampling technique for balancing the minority instances. They overcame the problems of under sampling and over sampling in their smart sampling algorithm, SMOTE, by generating synthetic samples of the minority instances to balance the training chunk. SMOTE is very effectively used for over sampling and many authors have used SMOTE in their hybrid resampling and have got accuracy improvement.

Later in their paper [25], the authors proposed a novel BD method that uses boundary conditions to improve the separation between the two classes in skewed data. The algorithm not only propagates the rare class instances but also the negative class that the classifier misclassified. Though this technique achieved higher precision and F1 measure, but the recall rate was lower than other comparative methods.

Another significant research **focused on algorithmic based methods** and a lot of researchers proposed ensemble methods for classifying imbalanced data streams.

H. Wang [3] proposed the first of the techniques using ensemble classifier on data chunks. The authors have used an ensemble of classifiers C4.5, RIPPER, Naive Bayes on sequential chunks of data streams. The classifiers are weighted as per their classification accuracy on the test data. Weights are assigned based on the distribution of the latest window. Only top K classifiers with the highest accuracy are kept. The technique then dynamically selects classifier for prediction without loss of accuracy.

Gao et. al [30] proposed SE (Sampling + Ensemble) a batch processing approach to balance the dataset. The authors have used an ensemble model and hybrid approach for sampling that includes under sampling of majority class and over sampling of minority class.

Chen et. al., [31] used a selective approach to balance the instances of minority class in an ensemble framework. They employ Mahalanobis distance to calculate proximity between the minority class of last batch and the current batch. Then the elements of the current batch are ordered based on the distance and first m elements are selected to maintain the ration between the classes. In the extension of their work further in [32] and [33], an ensemble of hypotheses is created on balanced data chunks using selective approach for selecting minority class data chunks. This approach allows the classifier to learn concepts from the imbalanced data.

Parallel research community also focused on **cost sensitive approach for imbalanced** data streams. Many authors have proposed the extension of the cost sensitive methods used for static data to the data streams.

Adel Ghazikhani et. al., [26] [34] in their research work on cost sensitive methods have focused on the use of ensemble of neural network classifiers with cost-sensitive objective function to deal with the problem of class imbalance in non-stationary data.

The cost-sensitive matrix can be a fixed or an adaptive. The adaptive cost matrix helps to deal with the non-stationary environment.

Quite a few frameworks have been suggested recently for **imbalanced data streams**.

S. Wang et. al., in [8] proposed the first of the online adaptive learner for imbalanced data streams. The authors proposed two methods, oversampling-based Online Bagging (OOB) and undersampling-based Online Bagging (UOB), to handle imbalanced data streams and make real-time predictions using resampling methods viz., OOB and UOB. The authors also propose an imbalance detector that dynamically calculates the imbalance ratio using a time decay factor to each instance that arrives. This factor reduces the effect of old data as new data arrives.

In [35] and [36], the authors propose a multi-objective sampling based ensemble model that builds hybrid online undersampling and oversampling online bagging models with the aim of increasing the accuracy of minority as well as majority class. The authors in [37] have extended their previous online ensemble model to multiclass classification for data streams.

In [38], B. Wang propose an online cost-sensitive ensemble learning framework that generalizes a batch of bagging and boosting based cost sensitive learning algorithm. Both these works have used a hybrid method combining data based sampling techniques with algorithmic based cost-sensitive methods.

The authors in [39] perform online learning of imbalanced data streams using Naïve Bayes classification using chunk based and incremental learning methods. The model is first created using a chunk of data. Initial training set is balanced by using k subsets of the size of minority class. Each k subset will be trained by k base models combined to form an ensemble and their average is taken. Model is updated in an online manner, such that only if a new instance is of minority it is used to update the ensemble. The technique considers only binary classes and availability of class labels and there is no support of drift.

The authors in [40] [41], propose a framework that work on datasets with very less labeled data. This framework considers imbalance data with drift. Only a few key samples are labeled, which can then be used for classification. The authors prove that it can be cost effective. The initial classifier is built on a chunk of data which is labelled by an expert. This initial classifier then predicts labels for next chunk. The predicted labels and the assigned labels are then used to update the model which is then used for further prediction. Ensemble of 10 classifiers is learned to perform classification. F-measure of classifier is used as a threshold to remove an underperforming classifier. The method works well for imbalanced data streams and claims to have improved the F-measure but though reduced it still requires an expert to label. It also assumes that the labels would be available for initial classification. Concept drift is detected based on the performance of the classifier. If it drops below a threshold then it is assumed that a drift has occurred. Concept drift detection methods can be improved as false alarms need to be identified.

Asim Roy in [42] used ensemble of Kohonen nets in two layers where the first layer gets a separation of class and in the second layer an ensemble of Kohonen submodel identify the presence of minority class. The technique can handle high dimensional high velocity data with the sub model computing in parallel on Apache Spark. However, high imbalance rate are not dealt with and second layer could be replaced by efficient classification techniques like decision trees or random forests.

Thus, the literature review covered discussion on various research papers related to the imbalanced classification. The next section discusses about the problem of imbalance in data streams. This problem is overlooked by most of the research papers discussed but it is observed to occur in many applications. A Spambase dataset is used as an illustrative example to depict the problem.

## 2.3 ANALYSIS OF IMBALANCE PROBLEM IN DATA STREAMS

In data streams, data can be processed in an online incremental manner or in a batch / chunk. When done in an online manner, the data instance is read one at a time and processed. The classifier then has to learn incrementally as the sample arrives. When done in batches / chunks, the input data is first collected in a batch and then a classifier is trained on this batch. In a streaming environment, the underlying distribution of data changes dynamically. As a result, there are several different possibilities in which the data can appear in a batch.

1. The data in a batch can be balanced having equal distribution of majority and minority class in a batch. This makes the classifier train on a balanced data giving good accuracy.
2. The majority class instances can be more than minority class in a batch. In this case, data based approaches to balance the data in the batch can be applied and train the classifier.
3. The majority class instances can be more than minority class in a batch like scenario 2 above. However, this distribution might be totally different from the distribution in the whole set. To apply suitable data based approach, knowledge of the distribution of majority and minority class is very important.
4. The batch might contain only majority class instances, making the classifier learn only the majority class. Various situations that may occur in this case are:
    a. There can be a scenario where the minority class instance is not seen for quite some time. The classifier keeps learning only the majority class instances. Now, later when a minority class instance arrives, it will be the only one that the classifier has seen in a long time. Applying data based approaches to balance the data with very few minority instances may lead to more problems than solution. Problems like adding meaningless instances or deleting important concepts.
    b. Minority class are misclassified. Generally, in imbalance the minority class instances are misclassified affecting many applications like medical diagnosis and fraud detection. In these applications, the cost of misclassifying minority class instance is very high.
    c. Minority class getting treated as noise. Since the classifier has not seen the minority class for a long time and the number of instances also being less, high probability that these will be treated as noise.
5. The minority class instances might have a different distribution of the classes in the batch. The batch contains only minority class instances or it contains more number of minority instances than the majority class.
6. Consider the Spambase dataset, where the ratio of majority to minority class for the whole dataset is **1.54 : 1**. For representation purpose, suppose that a batch of

13 instances is read at a time. The first two columns are two dimensions of the Spambase dataset and the last column represents the class label 0 and 1. The class label 0 represents the ham email whereas 1 represents the spam emails.

| 43 | 614 | 0 | | 71  | 201 | 0 | | 12  | 44   | 1 |
|----|-----|---|---|-----|-----|---|---|-----|------|---|
| 1  | 4   | 0 | | 26  | 77  | 0 | | 28  | 87   | 1 |
| 90 | 321 | 0 | | 6   | 54  | 0 | | 10  | 75   | 0 |
| 74 | 207 | 0 | | 11  | 48  | 0 | | 11  | 89   | 1 |
| 8  | 48  | 0 | | 5   | 44  | 0 | | 47  | 91   | 0 |
| 30 | 105 | 1 | | 20  | 965 | 0 | | 126 | 937  | 1 |
| 13 | 186 | 1 | | 110 | 309 | 0 | | 4   | 8    | 0 |
| 13 | 42  | 0 | | 10  | 61  | 0 | | 26  | 151  | 1 |
| 1  | 3   | 0 | | 10  | 58  | 0 | | 669 | 1402 | 1 |
| 3  | 19  | 0 | | 8   | 68  | 0 | | 38  | 508  | 1 |
| 7  | 97  | 0 | | 12  | 122 | 0 | | 13  | 146  | 0 |
| 26 | 693 | 0 | | 68  | 356 | 0 | | 11  | 113  | 1 |
| 82 | 197 | 1 | | 26  | 94  | 0 | | 5   | 70   | 0 |
| (a) | | | | (b) | | | | (c) | | |

**Figure 2-2: Different possibilities of distibution of data in Spambase dataset**

Different possibilities of distribution of data that may occur are shown in Figure 2-2 (a), (b) and (c).

Figure 2-2 (a) represents the possibility 2 where the majority class instances are more than the minority class. It also represents the possibility 3, where the ratio of majority to minority class is different (**3.33 : 1**) than that of whole dataset (**1.54 : 1**).

Figure 2-2 (b) represents the possibility 4 in which the batch consists of only the majority class. The classifier trains on this batch but the later batches can contain some minority class instances which the classifier has not seen earlier. It can treat the minority class as noise or can misclassify.

Figure 2-2 (c) depicts the possibility 5 where the ratio of majority to minority class is almost (**0.65 : 1**) again different than the overall ratio of (**1.54 : 1**).

Thus, it can be seen that the class distribution is different in each batch. Moreover, this distribution is not known to the learning algorithm a priori. Hence making any assumption of the distribution of data is not an option while dealing with imbalance data streams. As a result, existing methods have to be modified to be able to handle such variations.

## 2.4 GAPS IN IMBALANCED DATA STREAMS

Though, quite a research is focused towards the classification of imbalanced data streams which is observed from the literature survey discussed, there are still certain gaps which needs to be addressed. Imbalance data is handled very well in static data using the traditional approaches, but the situation is severe for data streams due to various reasons. After analysing the nature of data in imbalance data streams and doing a detailed literature review, it was noticed that there are several ways in which the data can appear as discussed Section 2.3. So far a general assumption of distribution and/or appearance of majority and minority class were made in the research so far. Most of the solutions proposed earlier assume that the distribution of classes is either known beforehand or the distribution remains same in all the batches. **There is a need to create frameworks** that can consider the various scenarios in which imbalance can occur. The frameworks should focus or handle situations as listed below:

1. **Long duration between minority instances** - Learning in data streams happen in an incremental manner. As a result, the next positive instance that the learning algorithm sees will be after quite a long time. This is because the positive class instances have **very few representations**. This **affects the learners' ability to learn the positive class** correctly [16]. The long duration between consecutive positive class instances makes it difficult to learn the positive class' separating boundary.

2. **Very few minority class instances** - Since the data instances are not available again after a learning algorithm has seen them, a positive class that arrives next might be the only one instance. There may not be a **prior occurrence of the minority class** for quite some time ahead. In such circumstances, neither the **undersampling of the majority class** nor the **oversampling with one positive class is possible** [43].

3. **Distribution not known** - The **distribution and arrival of data is not known apriori** in data streams. In a dynamically changing environment, the majority class seen in the current sample or window may actually be a minority class [8].

4. **Dynamic distribution** - The distribution of minority and majority class may interchange in a window [17]. A majority class may have very few

representations and treated as minority class in the current window. There won't be any representative of a class in the current window. Or both classes have a completely balanced distribution. This poses a challenge in determining the actual imbalance ratio.

5. **Class imbalance with concept drift** [16] [44] creates a severe effect on the learning algorithm. This is because of the very few positive instances which occur over a period in a stream makes it difficult to infer if it is noise or drift. The misclassification error can be due to noise or due to drift in concept. In the latter case then the drift must be detected correctly.

6. **Scarcely labeled data** - Almost all of the research papers discussed in section 2.2, assumed the availability of class labels and knowledge of the distribution of classes throughout the learning of the model, but in reality these class labels are not available at the time of learning or they may arrive after a delay. Learning in the absence of class labels requires another approach than the supervised ones discussed earlier.

Learning in data streams with imbalanced data is challenging majorly due to the dynamic distribution of classes in a batch. Applying sampling to balance the data then becomes a risky task which otherwise in traditional classification of imbalanced data solves the problem easily. Learning with imbalance data streams should happen in the absence of any apriori information about the distribution of data. Traditional algorithms as well as research done so far have failed to provide one solution to handle the scenarios. There is a need for learning algorithms that can improvise learning in classifying imbalanced data streams.

## 2.5 CONCEPT DRIFT DETECTION TECHNIQUES

Drift detection can be done by using different techniques which are briefly discussed below. These techniques include statistical based, window-based and ensemble based techniques. The techniques have been proposed by researchers and are now followed by many in the field of concept drift. The techniques along with their advantages, disadvantages and modifications are discussed in the sub section below:

## 2.5.1 Statistical based drift detection techniques

Most statistical based techniques are based on SPRT i.e. Sequential Probability Ratio Test that claims that given two distributions P0 and P1 for period w, if the underlying distribution shifts from P0 to P1 then the probability of observing elements from P1 should be higher than P0.

Different statistical methods include, CUSUM [13], Page Hinckley [14], DDM (Drift Detection Methods) [5] and EDDM (Early Drift Detection Methods) [45].

Table 2-2: Statistical based drift detection techniques

| Algorithm | Short Description | Advantages | Disadvantages |
|---|---|---|---|
| CUSUM (CUmulative SUM) [13] | CUSUM detects a drift when the mean of the input data changes significantly. The input is the residual of Kalman filter | - State-of-art drift detectors<br>- It is memoryless.<br>- It is well suitable for many applications in data streams. Also useful in anomaly detection in video streaming. | Accuracy depends on the parameters of the algorithm viz., threshold and magnitude of change. |
| Page Hinckley (PH) [14] | Page Hinckley detects drift when the average behavior of the input data changes. | - State-of-art drift detectors<br>- It is also memoryless.<br>- It is well suitable for signal processing applications. | Accuracy depends on the parameters of the algorithm viz., threshold and magnitude of change. |
| DDM (Drift Detection Methods) [5] | Compares the statistics of two windows where one window stores all the data and another window stores data from the start until increase in errors. When the prediction errors start increasing then alarm is raised | - Almost memoryless as windows are not stored in the main memory, only statistics is stored. | - Stores windows in secondary memory<br>- Suitable only for sudden drifts. Gradual drifts take longer time to detect and cause memory overflows. |
| EDDM (Early Drift Detection Methods) [45] | Same method as DDM but uses distance between the error rates to detect drifts | - Improvement over DDM | - more number of errors are required for calculations<br>- not suitable for imbalanced datasets |

|  |  |  | when drifts occur in minority class |
|---|---|---|---|

### 2.5.2 Window based drift detection techniques

Window based techniques monitor a window of instances instead of monitoring instances individually. These windows are sliding window of varying length or sizes that maintain statistical information. In all these techniques, deciding window size is very crucial. Large window size show higher accuracy but a drift might be present within the window itself whereas smaller window size can do better drift detection. These include VFDT, CVFDT, ADWIN and ADWIN2 which are discussed below:

Very Fast Decision Tree (VFDT) first proposed by Pedro et. al., in [46] is a type of decision tree that can replace traditional decision tree classification algorithms. It creates a tree as the data arrives and does not store the entire dataset. It uses Hoeffding Bound to select splitting attribute. It is seen that Hoeffding bound can also be applied to a sample of data stream and it gives the same splitting attribute as that of the entire data stream.

VFDT, however, is found to be not suitable for dynamic streams. They do not contain forgetting mechanism or relearning in the presence of drifts. VFDT were modified to make them capable of dealing with non-stationary data streams i.e. streams with drifts. The first is the Concept adapting Very Fast Decision Tree (CVFDT) [47], wherein a sliding window is used to store the tuples in the memory. Instead of creating a new model from scratch every time a new instance arrives, CVFDT just updates the existing decision tree after a seeing a certain number of instances. To decide which instances have to be forgotten, CVFDT increments the count of the most recent instance and decrements the count of the oldest instance in the window. The Hoeffding bounds are recomputed and if a better splitting attribute is found then a concept drift is believed to have occurred and a new subtree is constructed. The old subtree is replaced after sometime if CVFDT verifies that the new subtree generated is of higher quality.

A modification to CVFDT was proposed by Hoeglinger and Pears in [48], where in instead of the fixed window, a concept-based window was used. This window is updated based on the concepts instead of time of arrival. A decision tree is constructed from the leaves. The window stores the tuples until it becomes full. Any underutilized leaves are combined with the parent nodes. The tuples are then removed from the memory making space available for accepting new instances and allowing the tree to grow.

Liu et. al., in [49], proposed an improvement over CVFDT which identifies drifts very well but without focussing on the types of drift. E-CVFDT i.e. Efficient CVFDT proposed by authors uses a caching mechanism and treats types of drifts separately unlike CVFDT. It has shown to improve performance for gradual drifts but not so much for sudden drifts.

ADWIN – ADaptive sliding WINdow, first proposed by Bifet and Gavalda in [50] uses a sliding window of size W to detect and adapt to drifts. As the stream arrives it gets appended to the head of the window. When the average between two sub windows of W are large enough, then drift is said to have occurred and the window shrinks dropping the older data. The Hoeffding bounds were used to compute thresholds for cut off. But these bounds were computationally expensive. Also the memory requirements increased linearly with the window size.

ADWIN2 – an improvement to ADWIN was proposed by Bifet and Gavalda in [51] to overcome the resource requirements of ADWIN. ADWIN2 uses a memory-efficient data structure that stores a sliding window of length of logarithmic memory size and process the window in logarithmic update time. This allows the algorithm to update and store windows efficiently. The algorithm then evaluates a subset of windows to check for drift.

### 2.5.3 Ensemble based drift detection techniques

Ensemble improves performance of classifier by including the decision from not one but many classifiers. The ensembles work on chunk or batch based data streams to predict with better accuracy.

The Streaming Ensemble Algorithm, SEA [52] proposed by Street and Kim are the very first block based ensemble technique, where a fixed size ensemble are added to a model to classify. Size of ensemble is maintained by replacing weak performing classifiers. Drawback of SEA was that, it replaces the worst performing classifier with a classifier trained on recent batch, but this still leaves poor performing classifiers in the ensemble.

The drawback of SEA was improved in AWE Accuracy Weighted Ensemble [3] where the classifiers are selected by using mean square error to select n best classifiers and then replace all the poor performing classifiers with new ones. One drawback of AWE is the threshold that selects the best classifiers is difficult to set and in case of sudden drifts none of the classifier will be best and there will not be any classification at all.

AUE Accuracy Updates Ensemble proposed in [53] improved the AWE by updating the classifiers directly instead of updating the weights. The chunk size changes dynamically whenever there are no drifts detected. During this time, the classifiers train on a large chunk of data.

## 2.6 ADDITIONAL LITERATURE REVIEW ON CONCEPT DRIFT

Apart from the papers and techniques discussed above, additional literature review on the recent developments in the area of concept drifts is discussed in this section. The progress in the research of the **statistical based drift detection** includes the papers discussed below:

The authors in [54] propose a statistical method to detect drifts in unlabeled data streams. The drift detection happens as soon as a data point arrives. The difference in the mean for numerical data and the mode for categorical data from other data points is calculated and the martingale using Noob's inequality is compared. If the difference is greater than the martingale, the drift is said to be detected.

In [55], the authors proposed DDM-OCI an improvement over EDDM to deal with drifts in imbalanced datasets. The paper makes an assumption that, the drift occurs only in minority class and uses its recall to detect the drift. However, in reality this is not always the case. No assumption about occurrence of drift can be made. The authors, in [56], have devised a procedure that analyses the empirical loss in the learning of the model to detect drifts.

In [57], the authors propose a double weighted ensemble approach to handle the drift. The instances are weighted and the models in the ensemble are also weighted to remove the one with the lowest weight. The weights are assigned to the instances using the posterior probability given by Naïve Bayes classifier. The instances that are above a particular threshold are maintained in a buffer and are used for retraining the ensemble. In [58], the authors extended their previous model to a diversified dynamic online ensemble. Low and high diversity ensembles are maintained by replacing the Poison (1) by Poison (p) distribution.

In [59], Wang and Abraham proposed the Linear Four Rate (LFR) as an improvement over DDM-OCI where they consider the precision and recall of majority and minority class (four values) to detect drift. Drift is detected when any of these four values exceed a pre-defined threshold. This overcomes the drawback of DDM-OCI.

Further improvement is suggested by Yu and Abraham in their paper [60], where they proposed a Hierarchical Linear Four Rate (HLFR), a two layer structure to detect drifts. Layer one is used to monitor the four rates and the second layer validates to see if the drift detected is false or not.

Parallel research in the **window based drift detection** techniques are as discussed below:

Bifet et. al., in [61] propose the use of probabilistic adaptive windows to maintain old as well as recent data samples for detecting and adapting to a drift. This method uses two lazy learning methods viz., ADWIN to detect change and online bagging to add diversity to the ensemble.

In [62], Cassidy and Deviney, propose the use of an online Random forest that can evolve with time and can detect drift. The model learns incrementally on each new observation as it arrives. It also restructures the random forest depending on the temporal predictive performance.

In [63], Song et. al., proposed a Hoeffding tree based concept drift detection for sensor data. The proposed model uses a chi square hypothesis test between two adjacent windows to detect a change in the new block.

Losing et. al., in [64], proposed the use of Self Adjusting Memory SAM model for kNN classifier that used the adaptive window to detect heterogeneous types of drift in a stream. The model uses the STM and LTM biologically inspired models to store and process the stream as well as detect and to drifts. Reoccurring drift information is stored in the LTM and hence this model is successful in detecting these drifts.

Sun et. al., in [65] have proposed the use of adaptive windows for change detector and identifying different size blocks based on the drift for creating the ensemble. A drift is detected using the Hoeffding bound and after the detection of the drift the poorest performing classifier from the ensemble is retrained on the current window suggested by the adaptive window.

In [66], Adhikari et. al., proposed an amalgamation of Hoeffding Adaptive Trees, Drift Detection Method and Adaptive Window for classifying evolving data streams from an electric transmission system. The proposed amalgamation provides 98% accuracy for binary class and 94% accuracy for multiclass in less time and low memory usage.

Hassani in [67], proposed an adaptive window framework that uses a dynamic window to which focusses on the current information and discards the old data. The model uses a static pruning periods to update the information and also uses decaying mechanism to delete older data.

## 2.7 GAPS IN CONCEPT DRIFT

Considering the literature survey in Section 2.6, learning algorithms that work on non-stationary data streams still face with the following issues or challenges [14] [68]:

1. **Different types of changes** - The change can be sudden, gradual, incremental or reoccurring. Most classifiers adapt to certain types of drifts only. In real cases, there is no fixed pattern of drift occurrence. Any one or all different types of drifts can occur simultaneously. A single framework that can adapt to drift types of drifts needs to be designed.

2. **Most techniques depend upon the accuracy of the classifier to detect drifts**. If the classifier is weak, drift detection is not accurate and adaptation to the drift

might not happen at all or might be adapted to false drifts. Techniques depending either on Mean Square Error (MSE) or other metrics from confusion metrics are needed to be explored.

3. **Concept drift with imbalance** – as mentioned in section 2.4 above concept drift with imbalance is a severe problem [44]. Many researchers focus on these problems individually, but in real cases they can co-occur as depicted in the applications in section 1.2. There is a need to deal with the dual problem in one solution.

4. **Scarcely labeled data** - Individual research on concept drift discussed in section 2.6 also considers availability of labels. Learning in the **absence of labels** is still a challenge. There is a need for learning algorithms that can efficiently classify in the absence of fully labeled data.

## 2.8   SEMI-SUPERVISED LEARNING (SSL) APPROACHES

The majority of the learning algorithms for data streams assume that the class labels are available all the time and as a result all these algorithms have the information about the class distribution and the ratio of imbalance of the data. Hence, all the methods discussed in section 2.2 and 2.6 are supervised in nature and this makes the classification of imbalance data streams an easier task. However, in reality the class labels are not available all the time as cited in the examples of section 1.1. Hence, there is a need to design learning algorithms that can perform the task of classification using this scarcely labeled data.

Learning from labeled and unlabeled data is essentially the approach of semi-supervised learning. It consists of methods that utilize unlabeled data with labeled data to learn better models.

Data D in any semi-supervised learning problem, consists of a labeled instances $D_l$ and unlabeled instances $D_u$ where u >>> l.

$$D = D_l \cup D_u \quad (2.1)$$

$$D_l = \{(X_i, Y_i) \mid i = 1, 2, \dots, l\} \text{ and } D_u = \{(X_{l+j}) \mid j = 1, 2, \dots, u\}$$

where $X_l = \{x_1, x_2, \dots, x_l\}$ are data instances labeled with $Y_l = \{y_1, y_2, \dots, y_l\}$

$and\ X_u = \{x_{l+1}, x_{l+2}, \dots, x_{l+u}\}$ are unlabeled

Generally, the labeled data are scarce and unlabeled are abundant. Given this labeled and unlabeled data, the goal of any SSL approach is to learn a **function f** to predict the future unseen data. This is **inductive learning** and an SSL is generally an inductive learning approach. A **transductive learning** does not involve learning any function but predicting the labels of unlabeled data.

Some of the well-known SSL approaches are Self-training, Co-training, Graph-based training and Generative mixture models [69] [70] [71] [72].

### 2.8.1 Self Training

In this method, a classifier is trained on labeled data and used to predict the labels of unlabeled data. These predicted labels together with the labeled data are used to retrain the model and these steps are repeated. It is a wrapper class that can be used easily with any supervised learner. It works well in practice. These techniques are however susceptible to outliers. Also incorrect prediction of unlabeled data can be reinforced causing poor performance. With low percent of labeled data, classifier tends to overfit causing poor accuracy.

### 2.8.2 Co-Training

In co-training, a dataset is split into conditionally independent views and train separate classifiers on these views simultaneously using labeled data. Each classifier predicts the unlabeled data. The prediction of one classifier is then used to train other classifier. The learning process always selects the best classifier to predict unseen data. Search space of finding best classifier is reduced. Partitioning data into separate views is difficult. Classifiers must agree on both the views for labeled and unlabeled data for better performance.

### 2.8.3 Graph based training

In graph based training, a graph is created using labeled and unlabeled data where the data represent nodes and the similarity between them represents the edges. Different ways of graph construction are suggested like closest distance based, kNN based and domain knowledge based. Graph based training is useful in structured as well as unstructured data like text classifications, Natural language processing applications etc. These methods are quite efficient and improve results for low dimensional data. Smoothness assumptions of graph construction have adverse effect when the data is overlapping. Performance is poor when there is a high dimensional data.

### 2.8.4 Generative Mixture Models

These are the oldest SSL techniques. The technique uses an identifiable mixture distribution like Gaussian Mixture models to create mixture of components. Each component is then labeled using the labeled data. This labeling called as soft cluster labeling can happen using as low as one label per component. Expectation Maximization (EM) can be used to identify mixture components. Benefit of this method is it can be used with any probabilistic learning model. EM however is prone to local maxima. If the global maxima are far from local maxima then unlabeled data may hurt the learning. Wrong mixture model assumptions may lead to lower accuracy.

### 2.8.5 Cluster and Label approach

In this method, a clustering algorithm is used to cluster the labeled and unlabeled data. A supervised predictor is then used to learn from the labeled instances of each cluster. Unlabeled instances are then predicted using this predictor. It is possible to cluster even with a single labeled instance and a large amount of unlabeled data. This approach performs very well when the distribution can match with the true distribution of data. However, these methods are hard to analyze. If the decision boundaries are not clearly separated, the performance can be very poor.

The above mentioned semi-supervised learning approaches are compared in the following section.

### 2.9 COMPARISON OF SEMI-SUPERVISED LEARNING APPROACHES

In the following Table 2-3, the SSL approaches are compared w.r.t their advantages and disadvantages.

**Table 2-3: Comparison of Semi-Supervised Learning approaches**

| SSL Approaches | Advantages | Disadvantages |
|---|---|---|
| Self-Training | - can easily be used with any supervised learner.<br>- Works well in practice. | - Susceptible to outliers.<br>- Incorrect classification will be reinforced.<br>- With low percent of labeled data accuracy is not so good. |
| Co-Training | - Always the best classifier is selected to predict unseen data. Search space of finding best classifier is reduced. | - Partitioning data into separate views is difficult.<br>- Classifiers must agree on both the views for labeled and unlabeled data. |
| Graph based Training | - Efficient in data with low dimensions.<br>- Useful in structured as well as unstructured data | - Smoothness assumptions of graph construction have adverse effect when the data is overlapping.<br>- Performance is poor when there is a high dimensional data. |
| Generative Mixture Models | - Expectation Maximization can be used to identify mixture components.<br>- It can be used with any probabilistic learning model. | - EM is prone to local maxima. If a global maximum is far from local maxima then unlabeled data may hurt learning.<br>- Wrong mixture model assumptions may lead to lower accuracy. |
| Cluster and label approach | - Possible to cluster with even a single labeled instance and a large amount of unlabeled data. | - Methods are hard to analyze.<br>- If the decision boundaries are not clearly separated, the performance can be very poor. |

Semi-supervised learning approaches are being used for many different applications which were earlier done by either supervised or unsupervised methods. The traditional semi-supervised learning methods are modified and proposed by many researchers to make them efficient and beneficial for different applications. The following Table 2-4 lists some of these methods with their advantages, disadvantages and applications [70].

Review of Literature

Table 2-4: Summary of Semi-Supervised Learning

| SSL approach | Advantages | Disadvantages | Applications |
|---|---|---|---|
| Self-Training | - Implementation is simple and easy.<br>- Faster Computation. | - Errors or noise get reinforced<br>- Sensitive to imbalanced data<br>- Predefined terms difficult to estimate | - Chest X-Ray<br>- Image classification<br>- Text Classification |
| SETRED - SElf TRaining with EDiting [73] | - Strong Noise resistance.<br>- Can identify miss classification errors.<br>- Better results than standard self-learning. | - Sensitive to imbalanced data<br>- Error convergence can occur | - Hepatitis data classification<br>- Wine classifier |
| AAST - Auto Adjustable Self Training [74] | - Best Self Training method.<br>- All major drawbacks in standard self-training methods are taken care. | - Computation cost proportion to k. | - SONAR Mushroom Classification |
| Co-training | - Better confidence values. | - Both Assumptions need to be satisfied.<br>- Unsuitable for real world scenario.<br>- Views need to be carefully chosen. | - Image Segmentation. |
| Co-training with PAC Framework [75] | - High potential for improvement. | - Method preliminary in nature.<br>- Views need to be carefully chosen. | - Image Segmentation |
| COTRADE- Confident Co-training with Editing | - Reliable view communication.<br>- Best training procedure among other given methods. | - Performance lacks for some datasets. | - Advertisement Image Filtering Web-page classification |
| Graph Based SSL [76] | - Works very well for correct assumptions. | - A high number of parameters result in zero cross validation error. | - Person identification |
| Cluster then label [77] | - Lower training time.<br>- High accuracy. | - Misclassification errors exist. | - Pathology Image Classification |

43

| SSL approach | Advantages | Disadvantages | Applications |
|---|---|---|---|
| Cluster then label PSO [78] | - Uses best fitness function. | - Shows improvement but only for a limited size of data. | - Computer Aided Diagnosis (CAD) |
| Cluster then label Hungarian [79] | - Misclassification error avoided<br>- Accuracy proportional to size. | - Polynomial Time complexity. | - Utterance Classification |
| GMM [80] | - Identifies hidden relationships.<br>- Better cluster shapes than k-means. | - Optimizing the loss function is difficult.<br>- Initialization method sensitive. | - Pathology Image Classification<br>- Phonetic Classification |
| GMM-Online Learning [81] | - Robust learning technique. | - Shows improvement but only for a limited size of data. | - Channel Based authentication. |
| S3VM [82] | - High Dimensional Input Space.<br>- Sparse Document vectors.<br>- Aggressive feature selection. | - Label initialization problem.<br>- Initialization method sensitive. | - Text classification |
| TSVMSC [83] | - Spectral Clustering improves stability.<br>- No label initialization problem. | - Lacks rational cluster calculation approach | - Spectral Clustering |
| S3VM-us [84] | - Does not suffer from label initialization problem. | - Continued training hardly results in any further information exchange | - Optical Recognition of Handwritten Digits. |
| Generative Method- EM | - Works for paired input data<br>- Works for large dataset | - Suffers from noisy data.<br>- Performance hampered due to bias | - Text classification |
| Generative Method- M-VAE [85] | - Does not suffer from noise.<br>- Does not suffer from data imbalance. | - Classification penalty overhead | - RNA Classification |
| Generative Method- Discriminative [86] | - Bias Correction.<br>- Performs better for large data. | - Poor performance for small datasets<br>- Further study needed | - Text classification |

## 2.10 LITERATURE REVIEW ON SEMI-SUPERVISED LEARNING

Since majority of the proposed model assumed existence of class labels, their proposed frameworks are not suitable in the actual scenario where the class labels are available after a certain delay. Hence, the problem is treated as a problem of semi-supervised learning methods. Although many semi-supervised algorithms are developed for static datasets, there is very little research done data streams. Moreover, insufficient research exists for the problem of concept drift and data streams which is discussed below:

In [87], the authors propose a concurrent semi-supervised classification and clustering technique, to classify data streams with very less labeled data. The classification and clustering technique execute at the same time and help to improve each other's accuracy. The semi-supervised clustering considers the labeled and the unlabeled data and produces high quality clusters, whereas, the semi-supervised classification uses this clustering to classify. The authors have also used a fading model with exponential function that enables to detect the time gap between current time and arrival time of an instance. This aids in identifying new instances and forgetting old instances. Concept drift is detected using this instance selection.

The authors in [88], Jianjun et. al., have also proposed a clustering and classification approach to deal with binary classification of scarcely labeled imbalanced data samples. They use the self-training approach in classification using logistic regression and cluster and label approach in k-means clustering. And the choice of the technique is decided as: if no. of features is large, clustering is used and for extremely imbalanced class regression is performed.

Masud et. al., in [89], proposed a semi-supervised clustering approach using expectation-maximization mechanism to deal with the problem of scarcity of labeled samples. The objective is to minimize intra-cluster dispersion using EM. The objective function tries to create pure clusters. The label propagation is performed using Gaussian affinity equation to create affinity matrix based on similarity of data samples. Classification is performed using ensemble of classifier that use inductive label propagation technique.

# Review of Literature

In [90] and [91], the authors have proposed a novel semi-supervised learning approach that uses computational geometry for dealing with the problem of extreme verification latency in initially labeled streaming environment. The authors have proposed a solution to the problem where labeled data is available only initially i.e. initially labeled environment (ILE) and rest all other times the data is unlabeled. Their framework COMPOSE (Compacted Polytope Sample Extraction) selects labels from the current time step where label is provided and uses these labels to label the unlabeled data at all other times.

In [92], the authors propose an online self-training model to learn the classifiers from unlabelled instances. The model is an ensemble of classifiers which is trained using the labeled instances in each chunk. The diversity of the ensemble is maintained by bootstrapping the labeled data. An unlabeled batch is then given to the ensemble to predict, the classification error rate is calculated using the labeled data. If the error rate is less than a threshold, then class of unlabeled data is predicted using majority vote from the ensemble. The framework uses a modified self-training approach by considering the performance of weak classifiers as well unlike regular self-training models. The proposed model performs better than the one proposed in [89].

Hosseini et. al., in [93] proposed an ensemble of cluster based classifiers to classify non-stationary data streams. In the ensemble each classifier is a pool of clusters based on a single concept. Clusters are formed using labeled and unlabeled instances. When the labels are revealed they are used to update the classifiers in the pool. The model has proved to outperform model in [89] and is an improvement as it is also able to detect recurring concepts.

Some of the very recent researches towards the application of semi-supervised methods in classification of data streams which are discussed as below.

In [94], the authors proposed a semi-supervised ensemble of classifiers to learn from time-varying data streams. The authors have used an online variant of Random Forest called CoForest that which is similar to co-training of semi-supervised learning and employs bootstrapping by assigning weights to instances. An Adaptive Windowing technique (ADWIN2) is also used to detect and adapt to concept drift.

In [77], Peikari et. al., proposed a cluster-then-label method to classify pathology images. Clustering was applied to identify high-density regions in the data space using the labeled and unlabeled data. A distance matrix is derived using the distance between the labeled and unlabeled core radii points. Labels are propagated to unlabeled radii points using this distance matrix. A supervised training algorithm SVM is then learned on these clusters to find the cluster boundary in the sparse region. This method has seen to considerably improve the performance in the classification of pathology data.

Some of the researchers proposed frameworks for classifying data streams that are relevant and related to this research problem. A comparison of these classification frameworks suggesting their methods and drawbacks is listed in Table 2-5 below.

Table 2-5: Comparison of existing Semi-supervised frameworks

| Algorithm | Method discussion | Drawbacks |
|---|---|---|
| REDLLA - REcurring concept Drifts and Limited LAbeled data [95] | - Deals with recurring drifts and data with limited labeled.<br>-A semi-supervised classification technique that uses decision tree technique.<br>- The leaf of decision tree uses K-means to produce clusters for concepts.<br>- The labels of the majority class are used to label the unlabeled data. | -Majority instance labeling gives preference to the majority class instances.<br>-Performs poorly for imbalanced data stream. |
| ReaSC - Realistic Stream Classifier [89] | -Uses constraint-based, semi-supervised clustering to create K clusters.<br>-Cluster and label approach is used to label unlabeled clusters using inductive label propagation.<br>-Drift detection is uninformed.<br>-Model updates when a new class definition is seen | -Models are updated purely on last trained data. This is sensitive to noise.<br>-Cannot cope with recurrent labels.<br>-Does not deal with imbalance data streams. |
| SAND - Semi-Supervised Adaptive Novel Class Detection and Classification over Data Stream [96] | -Uses an ensemble of clustering based classifier that uses classifiers confidence scores to select instances for labeling.<br>-Uses an adaptive window to detect drifts and novel class detection which is done using cohesion among data. | -Adaptive window employed by SAND is computationally expensive due to exhaustive invocation of the change detection module. |

Review of Literature

| | | |
|---|---|---|
| SPASC - Semi Supervised Pool and Accuracy based Stream Classification [93] | -A pool of classifiers wherein each classifier determines a specific concept is maintained. The concepts are created using semi-supervised clusters. Each concept is assigned the class of the max labeled instances.<br>-A classifier is added to the pool when a new batch is read and the previous classifiers are unable to classify accurately.<br>-The number of classifiers in the pool is kept constant by removing the classifier with lowest weight.<br>-Drift detection is uninformed. | -Majority class instances in a cluster suppress the minority classes. As a result in imbalanced data streams, the minority class clusters are hardly created.<br>-This affects the minority class recall.<br>-It assumes Gaussian distribution by default which is not true for large datasets.<br>-Due to uninformed drift detection, unnecessary training of model is done on the misclassified instances leading to poor performance. |

## 2.11 ANALYSIS OF SEMI-SUPERVISED LEARNING

From the literature survey discussed in section 2.10 above, it is observed that Semi-Supervised Learning techniques can improve the performance considerably. It is also observed that, most of the researchers have focused on the classification of data streams that contain labeled and unlabeled data both. However, to the best of knowledge, none of the researchers has applied semi-supervised learning to deal with the problem of imbalance in data streams. Considering the background study of semi-supervised learning and the literature survey semi-supervised learning it can be concluded that:

1. Semi-supervised learning are suitable for problems which consists of labeled and unlabeled data as they are able to exploit the advantages of available unlabeled data.

2. If the percent of labeled data available is very low, then cluster and label approach or Gaussian mixture models can prove beneficial than self-training and co-training methods.

3. Semi-supervised learning techniques to solve the problem of imbalance data streams has not been applied as of now. Employing clustering techniques will be beneficial for imbalanced data streams where the minority class are very less.

Care has to be taken to ensure the majority class do not create larger and sparser clusters.

## 2.12 SUMMARY

This chapter discussed about the various techniques and approaches for dealing with imbalanced data and concept drifts. It also discusses about the existing semi-supervised learning approaches. The chapter also provides a detailed literature review on each of the respective topics of imbalance classification, concept drifts, and semi-supervised learning. Gaps identified and analyses of the approaches are also explained for respective topics. In the next chapter, reformulation of research problem, research objectives, research scope and contributions are discussed.

# Chapter 3

Research Problem, Objectives and Design

# 3. RESEARCH PROBLEM, OBJECTIVES AND DESIGN

A detail study of data streams, imbalance data streams, analysing the problem of imbalance in data streams, concept drift and literature review on each respective topic, it is observed that, a lot of research exists that deal with the problems of imbalanced data streams and concept drifts individually. There are a few researches that address the issues of imbalanced and concept drifts to some extent. But, they assume the distribution of data is same throughout the time the data is read and that the data is labeled all the time. As a result, it is convenient to use supervised learning algorithms for classification. In real time however, the distribution of data changes, the ratio of majority to minority class is not the same always and fully labeled data may not be available all the time. In the absence of labeled data availability, these learning algorithms fail to provide efficient classification accuracy. Hence, there is a need to design a framework that can improvise the classification of imbalanced data streams with concept drift in the absence of fully labeled data. This leads to redefining the problem statement of Section 1.7.

The research problem is then stated as follows:

**"To analyze and evaluate the problem of imbalance classes in streaming data and design a model for learning to perform classification addressing the issues of occurrence of imbalance, class label unavailability and imbalance data with concept drift."**

The research aims at recommending a solution framework to solve the problems occurring in a data stream. The research objectives and scope of the research are explained in the section 3.1 and 3.2 below.

## 3.1 RESEARCH OBJECTIVES

As stated earlier in section 1.7, the aim of the research is to design a framework that is able to classify imbalanced and non-stationary data streams. The challenges and gaps aided in defining the research objectives. Considering the learning requirements of data streams discussed in section 1.4.3, the challenges and gaps while dealing with

# Research Problem, Objectives and Design

imbalanced data streams discussed in section 1.5.1, 2.3 and 2.4 and also those while dealing with concept drift discussed in section 1.6.2 and 2.7, the research objectives are:

## 1. Processing of data streams online

Since data is continuously generated at constant time interval $t_i$ in data streams, the framework should be able to process the data stream as it arrives. In this regard, the objectives are:

a. To process data at the time of arrival. Unlike a regular learning model, the entire dataset will not be available at any time, the data cannot be stored permanently and the data cannot be visited as and when required.

b. To devise a method to read and process input data, allowing the framework to overcome the problem of loss of data due to its fast arrival speed.

c. To analyze the advantages of chunk based and incremental learning and select appropriate learning suitable for the research problem.

## 2. Imbalanced class distributions

Data streams are a dynamic flow of data. The distribution of data may change dynamically. This dynamic change also affects the ratio of imbalance. The research objectives in this regard are:

a. To recommend appropriate balancing technique (data based or algorithmic based) for data streams.

b. To perform learning in the presence of imbalanced data. While learning to classify data streams, neither the assumptions about the nature and type of distribution nor about the ratio of class distribution can be made.

c. To improve the recall of minority class. Irrespective of the imbalance ratio the objective of the framework is to improvise learning of the minority class.

## 3. Inherent concept drift

The data stream under consideration is assumed to be non-stationary and hence the underlying distribution of the data may change overtime. The objectives of the research in this context are:

a. To devise a method that is able detect a drift as soon as it occurs. It is important for a learning algorithm working on a data stream to be able to detect a drift as soon as it occurs.

b. To recommend a technique that can detect any type of data. No assumption about the occurrence of the type of drift should be made. As the nature and type of drifts is not known a priori, hence it is required to have a generic drift detection technique as opposed to a specific one.

c. To recommend a method to adapt to the drift once detected. Adapting to a drift requires for the model to be retrained on the data where drift has occurred. The objective then is to be able to select correct data for retraining the learning model to make it suitable for the changed concept and improving the accuracy of learning.

d. To evaluate the performance of the model by evaluating if the recommended method is able to improve the accuracy irrespective of the drift occurring in majority or minority class.

## 4. Dealing with unavailability of labels

Considering that the labeling of data is a cumbersome and time consuming process, one of the major objectives of the research is

a. To recommend a model that can perform learning in the absence of fully labeled data. Since in data streams, the labeled data is scarce and unlabeled data is abundant, it is required that the learning model exploits the advantages of using the unlabeled data together with the labeled data to learn the model.

## 3.2 SCOPE OF THE RESEARCH

The proposed model is able to address the following concerns:

Research Problem, Objectives and Design

1. **Processing data as it appears:** The designed framework will process the data as soon as it appears, some temporary storage as a buffer can be used but the data will not be revisited again and again for training purpose; as a result there is very little or no scope for pre-processing the data stream.

2. **Binary class imbalance:** The study and analysis in this research is limited to imbalance of binary classes only due to the lack of suitable literature on multiple classes. Though some of the related literature review has designed a framework that can be extended for multiple classes as well, however, they all assume the class ratio is known a priori and the distribution of classes in each batch is same as that of the whole batch. Multi-class imbalance problem in data streams can be done with assumptions of the data like distribution of each class instances in a batch, occurrence of all the class instances in a batch, information about the minority class labels, multi-minority or multi-majority type of problem, drift occurring in majority or minority etc.

3. **Dealing with numerical data:** Currently in this research, datasets with numeric attributes are considered. Euclidean distance is used as a measure of similarity between objects. As a future scope, attributes having categorical data types can be considered.

## 3.3 RESEARCH DESIGN

The research methodology followed is applied analytical and experimental by analysing the performance of the model proposed to solve the research problem. As explained in section 1.8, the research design involves first designing a hybrid ensemble model for classifying imbalanced datasets, HECMI, which deals with imbalanced data using sampling for balancing the data and in-built algorithms to classify the data. The novelty is in the way the data is partitioned and selected for oversampling. This model improves the performance for imbalanced datasets and is at par with the standard algorithms for the balanced datasets.

HECMI is a static model that worked on static data which involved visiting data more than once. A second model, Concept drift Detection and Adaptation model for Classifying Imbalanced data streams, CDACI, is an incremental learning model that is able to detect drift in the data and adapt to it. It uses sampling to balance the data

Research Problem, Objectives and Design

that is visible in a window of instances. It also uses in-built classifiers to perform classification. It is able to improve the classification by detecting and adapting to drifts.

These two models are then analyzed to identify the benefits and drawbacks in them. Using some of the best design factors from these two models, an adaptable semi-supervised learning algorithm that is able to overcome the limitations of HECMI and CDACI and also improve the performance of classifier to a considerable extent called ASSCCMI is designed.

Each of these models, viz., HECMI, CDACI and ASSCCMI, are evaluated against different benchmark datasets. HECMI and CDACI are evaluated on selective datasets suitable for their problem statement from these 7. The performance of HECMI and CDACI are compared against standard supervised classification techniques whereas ASSCCMI is compared with standard stream classification and existing semi-supervised learning frameworks. Test cases and hypotheses are used for evaluating ASSCCMI and prove its improvement.

## 3.4 CONTRIBUTIONS OF THE RESEARCH

This research focuses on designing a framework for classifying imbalance data streams with concept drifts. The major contributions of this research are in improvising the overall accuracy of the model in classifying the data streams in the absence of fully labeled data, improvising the minority class recall in imbalanced data streams and detecting and adapting to a drift in the data streams.

**Major Contributions:**

1. Designing an Adaptive Semi-Supervised Clustering-based Classification Model for scarcely labeled Imbalanced data streams (ASSCCMI). The designing of the model involved

    a. Designing a Semi-Supervised clustering method using Expectation-Maximization technique (SSL_EM) to cluster labeled and unlabeled data.

    b. Recommending cluster purity check technique to verify the cohesiveness of clusters and creating homogenous clusters if the clusters created by the SSL_EM techniques are not pure.

# Research Problem, Objectives and Design

    c. Designing a similarity based label propagation technique to label unlabeled data based on its similarity to the labeled data belonging to the same cluster.

    d. Designing a novel technique to merge micro clusters for balancing the number of micro clusters. This is required when the number of micro clusters exceed a predefined number.

    e. Recommending a window based statistical approach to detect concept drift in the data streams and adapt the classifier to improve performance.

2. Designing a Concept drift Detection and Adaptation Model for Imbalanced data streams (CDACI) with fully labeled data. The model recommends a window based approach for dealing with the data streams and comparing adjacent windows for detecting drifts.

3. Designing a Hybrid Ensemble Model for Classifying Imbalanced data (HECMI). The model recommends use of ensemble models and sampling techniques to classify imbalanced static data.

**Minor Contributions:**

1. Designing an input module to read data from data streams and provide to the ASSCCMI model.

2. Designing an ensemble classification model with diverse classifiers to evaluate the proposed Semi-Supervised clustering method using Expectation-Maximization (SSL_EM) and compare its performance.

3. Designing standard stream classification techniques to evaluate the performance of the recommended ASSCCMI approach against the performance of these approaches.

## 3.5 SUMMARY

In this chapter, detailed problem formulation, research objectives and scope, research and contributions of the research are discussed. The next chapter discusses about a proposed model HECMI that deals with the problem of imbalanced data. The results, analysis and limitations of the model are also discussed in the following chapter.

# Chapter 4

Hybrid Ensemble Model for Imbalanced data - HECMI

# 4. HYBRID ENSEMBLE MODEL FOR IMBALANCED DATA - HECMI

Classifying imbalanced data can be done either by applying data-based approaches or by algorithmic-based approaches as discussed in chapter 2, section 2.1. A hybrid of both the approaches can be considered to improve the classification accuracy. In the cost-sensitive approach of algorithmic based approaches, deciding the misclassification class for a dataset is a major task. It requires the domain knowledge and an expert's opinion to assign the cost. In this chapter, a hybrid model is designed to classify imbalanced data. The designed hybrid model, HECMI, is a combination of data-based and algorithmic-based approach. It combines the sampling approach to balance the dataset and ensemble approach to classify the dataset. The model deals with the problem of imbalanced data giving a better accuracy and recall. However, there are certain assumptions about the data and the nature of the problem that this model considers. These assumptions are as follows:

1. It is a supervised learning model hence, the class labels for all the classes i.e. majority and minority are known apriori.

2. The data is static; the whole data can fit in main memory. The data can be read as and when necessary.

Refer [97] for a technical paper on this model. This chapter is organized as follows:

HECMI Design → Experimentation → Result Analysis → Limitations → Analysing HECMI

**Figure 4-1: HECMI Chapter Organization**

## 4.1 HECMI - THE MODEL

HECMI is an ensemble model that combines different traditional classification techniques to classify an imbalanced dataset. For balancing the data, HECMI employs a novel data partition method that creates a training part and testing part from each partition. A model is trained on each training part and tested on testing part. The misclassified instances from the testing part are then added to the training part of the next partition. This process is continued until no further partitions remain. For training, HECMI uses traditional classification approaches like Logistic Regression (LR), Linear Discriminant Analysis (LDA), K Nearest Neighbors (KNN) and Support Vector Machine (SVM). The algorithm with a better performance is then selected as a base classifier for the ensemble. A majority voting of the prediction from all the classifiers in the ensemble gives the final class for unseen data. The HECMI algorithm is shown in Table 4-1 below.

**Table 4-1: HECMI Algorithm**

| Algorithm : HECMI |
|---|
| **Input:** <br> 1. D: Original Dataset <br> 2. $C_i$: Base classifiers <br> **Output:** <br> 1. Ensemble classifier E <br> 2. Predicted class label $c$ |
| **Steps:** <br> 1. Read ← original dataset D <br> 2. Select the base classifier. <br> 3. Base classifier ← classifier with max cross validation score on D. <br> 4. Split(D) into training and testing part <br> 5. Training Part ← Split(D)　　//70% of part D <br> 6. Testing Part ← Split(D)　　//30% of part D <br> 7. Split training set into N data partitions. <br> 8. For each data partition in i = 1 to N do <br>　　a. Split partition into training and testing part i <br>　　b. Create a Classifier $C_i$ on the training part i <br>　　c. Evaluate $C_i$ on testing part i for each class <br>　　d. E ← Add classifier to the ensemble <br>　　e. Modify data part i+1 <br>　　　　i. Oversample the class with the least recall value <br>　　　　ii. Collect instances of classes with recall less than the threshold <br>　　　　iii. Append the instances from i & ii to data part (i+1). <br>　　f. Move to next iteration (i+1) <br> 9. Read data x <br> 10. Get class label for x: $c_i$ = $C_i$(x). <br> 11. Class with majority votes is the final label. <br> 12. Final c = argmax $\sum_{i=1}^{N} c_i$ <br> 13. Return $c_i$ |

The model begins with collecting the dataset for training and creating the ensemble. The dataset is stored in the main memory and can be visited as many times as it is required. The design of the model is shown in Figure 4-2.

### 4.1.1 Base classifier selection

A base classifier is selected from various traditional classifiers like Linear Discriminant Analysis (LDA), Naive Bayes, Logistic Regression (LR), Support Vector Machine (SVM) and K Nearest Neighbors (KNN). The base classifier is selected from the one having high accuracy for the dataset. The benefit of this approach is that diversity will be maintained in the ensemble and it will help to increase the accuracy and recall as compared to other classifiers. For Spambase and Electricity dataset Logistic Regression had the highest accuracy, whereas for SEA1 and SEA2 Linear Discriminant Analysis had highest accuracy and recall for minority class. For Hyperplane1 and Hyperplane2 almost all classifiers had similar accuracy.

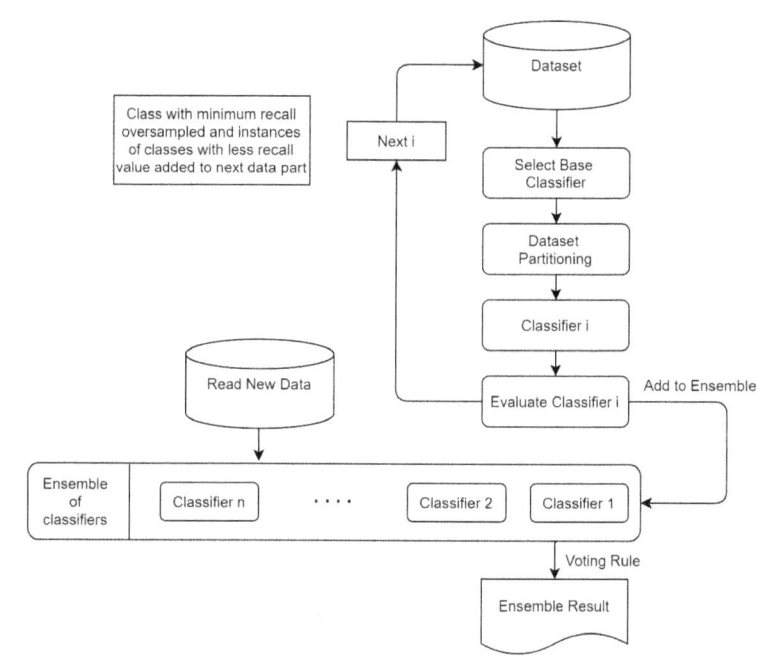

**Figure 4-2: Design of HECMI model**

### 4.1.2 Dataset Partitioning

A three step approach is used to classify imbalanced dataset as shown in Figure 4-3. First, the original data set is partitioned into two parts: Training and Testing. Second, the training part is then divided into N parts called Data Part i (i = 1 to N). Third, the data part is further partitioned into 2 sets Training set 1 and Testing set 1. A classifier is trained on the training part 1 and tested on the testing part1.

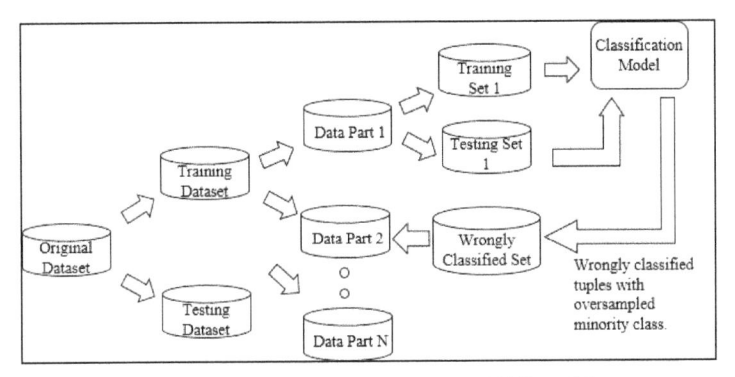

Figure 4-3: Dataset partitioning of HECMI model

### 4.1.3 Ensemble of classifiers

The ensemble creation is an iterative process that is done once the base classifier is selected. The ensemble consists of total N classifiers where N = number of classes in the dataset. Each classifier is trained on the training set and tested on the testing set. Along with the misclassified instances of minority class, the instances with recall below a particular threshold (< 70%) are oversampled using SMOTE [24] to balance the dataset and added to the data part2. This data part is again divided into a training set and testing set and a second classifier trained is added to the ensemble.

### 4.1.4 Ensemble Result

A majority voting technique is used for the final prediction. Each classifier predicts a class label $c_i$ for the unseen data. The class label with the maximum vote is given as the final class label for the unseen data.

$$Final\ c = argmax \sum_{i=1}^{N} c_i \qquad (4.1)$$

## 4.2 DATASETS FOR HECMI

HECMI is evaluated on real as well as synthetic datasets (refer Section 7.2 for further details). The datasets have varying imbalance. The imbalance ratio (IR) is calculated by taking the ratio of majority class instances to minority class instances. The datasets are briefly described with their properties in the Table 4-2 below:

Table 4-2: Dataset properties for HECMI

| Dataset Name | Real / Synthetically generated | Total No. of instances | No. of attributes | Imbalance ratio (IR) |
|---|---|---|---|---|
| Hyperplane1 [98] | Synthetic | 100000 | 10 | 1.01 |
| Hyperplane2 [98] | Synthetic | 100000 | 10 | 1.00 |
| SEA1 [52] | Synthetic | 100000 | 3 | 1.79 |
| SEA2 [52] | Synthetic | 100000 | 3 | 1.32 |
| Electricity [99] | Real | 45312 | 7 | 1.36 |
| Spam base [100] | Real | 4601 | 57 | 1.54 |

## 4.3 RESULTS AND CONCLUSION FOR HECMI

The performance of HECMI is compared with standard classification techniques like Logistic Regression (LR), Linear Discriminant Analysis (LDA) and Support Vector Machine (SVM). As compared to other classification techniques, these techniques gave the best results during validation. Hence, HECMI results were compared with these three classification techniques. The following Table 4-3 compares the performance of HECMI with these algorithms for the dataset.

Table 4-3: Comparison of Accuracy and Recall for HECMI model

| Datasets | Performance | LR | LDA | SVM | HECMI |
|---|---|---|---|---|---|
| Spambase | Accuracy | 0.93 | 0.89 | 0.70 | **0.93** |
| | Recall | 0.91 | 0.79 | 0.44 | **0.94** |
| Electricity | Accuracy | **0.75** | 0.73 | 0.73 | 0.72 |
| | Recall | 0.58 | 0.54 | 0.55 | **0.63** |
| SEA1 | Accuracy | 0.84 | 0.84 | 0.84 | **0.84** |
| | Recall | 0.72 | 0.73 | 0.70 | **0.73** |
| SEA2 | Accuracy | 0.83 | 0.83 | 0.83 | 0.83 |
| | Recall | 0.71 | 0.72 | 0.70 | **0.73** |
| Hyperplane1 | Accuracy | 0.76 | 0.76 | 0.76 | 0.76 |
| | Recall | 0.76 | 0.76 | 0.76 | 0.76 |

| Hyperplane2 | Accuracy | 0.85 | 0.85 | 0.85 | 0.85 |
|---|---|---|---|---|---|
| | Recall | 0.85 | 0.85 | 0.85 | 0.85 |

HECMI selects the best of the classifier based on its performance on a particular dataset. The dataset partitioning method helps to improve the classification and is also evident from the Table 4-3. HECMI also improves the minority class recall for imbalanced datasets viz., Spambase, Electricity, SEA1 and SEA2. For balanced datasets the performance of HECMI is at par to the standard classification algorithms. However, HECMI has certain limitations when it comes to dealing with data streams and concept drifts which are listed below in section 4.4.

## 4.4 LIMITATIONS OF HECMI

HECMI is a basic model designed to classify imbalanced data. Though it was able to successfully classify imbalanced data for binary as well as multiclass, it had certain limitations.

1. **Assumes whole dataset is available:** The dataset is static, so the classifier is able to see the entire data and scan as and when it is required. The whole dataset is available at any time. This contradicts the first requirement of data stream learning that data is processed only once.

2. **Memory requirements are huge:** Since the whole dataset resides in main memory, the model requires enough memory to store the complete dataset in the memory. This again contradicts the limited memory requirement of data stream learning. For small dataset it is feasible to store the complete dataset in memory but for larger datasets the size of memory required is huge.

3. **Assumes fully labeled data:** HECMI is a supervised classification technique and assumes that the data is fully labeled at all times. However, the data streams face a problem of unavailability of labels.

4. **Hybrid approach in case of binary class doesn't show considerable improvement in recall:** HECMI uses the data based approach of oversampling and algorithmic based approach of ensemble to deal with imbalanced data. It works best for multiclass and larger datasets as is observed in results in [97].

With binary datasets however, the capability of HECMI is not explored to a beneficial extent and there is improvement for only a few datasets.

5. **HECMI cannot detect concept drifts:** HECMI was designed to classify in the presence of imbalanced static data. Drifts do not occur in static data. HECMI was not modelled to detect drifts. Hence it does not detect concept drifts.

## 4.5 SUMMARY

HECMI assumes the data is static and hence it is able to scan the dataset more than once for selecting the best classifier, for partitioning the dataset and for training the model. It also assumes the data is fully labeled and hence it is a supervised learning model. Due to its inability to deal with non-stationary data streams, it is not able to adapt to drifts in the data. An explicit drift detecting model that can work with data streams can overcome the limitations of HECMI.

A novel concept drift handling model CDACI is proposed in the next chapter. This model is a supervised model designed to classify imbalanced data streams. The model overcomes the limitations of HECMI, by detecting drifts and adapting to them in the presence of imbalanced data streams.

# Chapter 5

Concept Drift Detection and Adaptation Model - CDACI

# 5. CONCEPT DRIFT DETECTION AND ADAPTATION MODEL - CDACI

The HECMI model discussed in the previous chapter deals with the problem of imbalanced data. It assumes static data and that the whole dataset can be visited as and when required. However, this contradicts the very first requirement of learning for data stream which states that a data is processed at the most once. Also HECMI is not designed to deal with concept drift as it works on static data.

In this chapter, a window based concept drift detection model Concept drift Detection and Adaptation model, (CDACI), for Classifying Imbalanced data streams, is designed that deals with the problem of detecting and adapting to the drift in the data in an imbalanced data stream. An existing classifier like Naïve Bayes or Logistic Regression is used to classify the imbalanced data stream. The model then uses cosine similarity between actual and predicted labels to detect a drift. The classifier then learns from the instances of the window where the drift is detected forgetting the previous concepts. CDACI deals with the problem of imbalanced data as well as concept drift. However, it does have certain assumptions about the nature of the problem. These assumptions are as follows:

1. It is a supervised learning model hence, the class labels for all the classes i.e. majority class and minority class are known a priori.

2. The ratio of the majority and minority class is same in all windows and matches the overall ratio of the whole dataset.

3. Imbalanced data are first balanced hence, irrespective of the type of class, drifts can occur in majority or minority class.

Refer [101] for a technical paper on this model. This chapter is organized as follows:

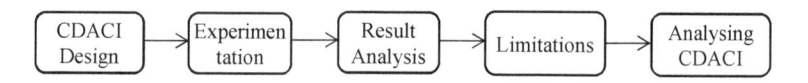

CDACI Design → Experimentation → Result Analysis → Limitations → Analysing CDACI

**Figure 5-1: CDACI Chapter Organization**

## 5.1 CDACI – THE MODEL

CDACI is an incremental model that learns on the data stream as soon as they arrive. The data is read in a batch of fixed size windows. The size of window is decided experimentally. The window size with more than average accuracy for each dataset is selected after thorough experimentation (refer section 5.2.2). The model consists of primarily three modules viz., the class imbalance detection, concept drift detection and concept drift adaptation. The model design is as shown in the Figure 5-2.

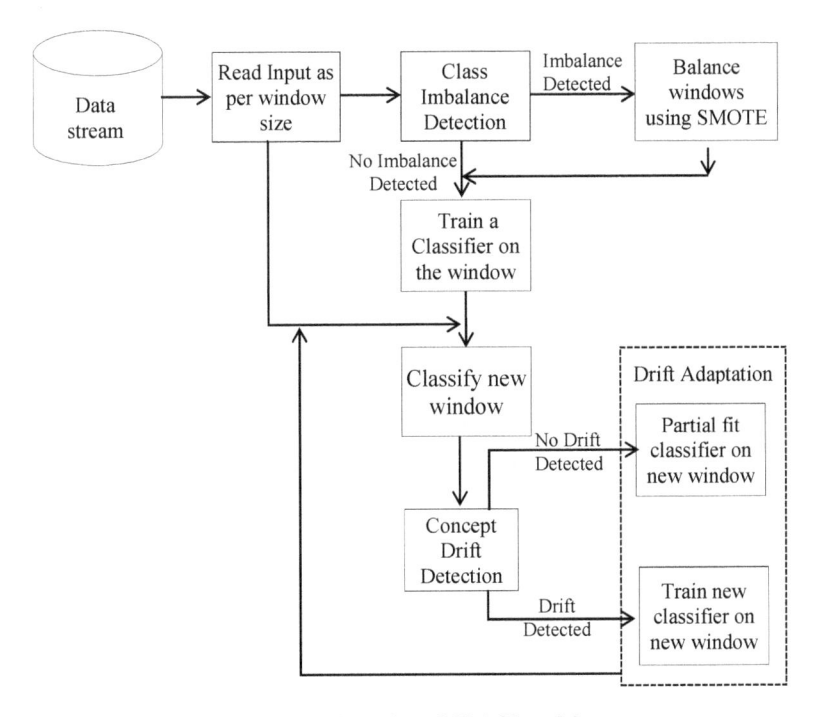

**Figure 5-2: Design of CDACI model**

The different processes followed in the design are explained in detail below:

### 5.1.1 Read the Input stream

As mentioned earlier, the input is read in fixed size windows. The incoming stream is split into equal sized chunks of predefined size. The window size for each dataset is decided experimentally. The size of the window is kept to be neither too small nor too large.

For large window size, there is a chance of missing out potential drifts and for too small a window size; the classifier will raise many false alarms.

### 5.1.2 Class Imbalance detection

For each window that is being read, the imbalance detection process calculates the imbalance ratio of majority to minority class. If this ratio is more than 1.2 times, then the window is balanced using SMOTE [24]. SMOTE balances the window by oversampling the minority instances. Oversampling is done by generating synthetic samples using k-nearest neighbours.

### 5.1.3 Train a classifier

A classifier is now trained on the balanced window from scratch. CDACI uses logistic regression as a default classifier. The classifier learns from the labeled data of the window. Once the classifier is trained a new window is read.

### 5.1.4 Classify new window

Now, the trained classifier is used to predict on the next window that is read. The performance of the classifier is evaluated against the actual class values. The accuracy and recall values are evaluated. The drift detection module is then invoked.

### 5.1.5 Concept Drift Detection

This module first calculates similarity between the actual value and the predicted value using the cosine similarity measure. The cosine similarity is a distance measure though it is called an angular distance. The cosine similarity measures the angle between the unit vectors and these angle measures provide a linear distance between the data points [102]. The range of the value is [-1, 1]. Higher the similarity value, closer are the points. The cosine similarity is calculated by:

$$similarity = \cos(\theta) = \frac{A \cdot B}{\|A\|\|B\|} = \frac{\sum_{i=1}^{n} A_i B_i}{\sqrt{\sum_{i=1}^{n} A_i^2} \sqrt{\sum_{i=1}^{n} B_i^2}} \quad (5.1)$$

Where A and B are the vectors amongst whom the similarity is to be identified.

In CDACI, A and B are the actual and predicted values returned by the classifier for the window. The similarity measure between the accuracy values; actual and predicted, is calculated. If it is more than a predefined threshold then a drift is said to be detected, else drift is not detected.

### 5.1.6 Concept Drift Adaptation

On detection of the drift, the classifier is adapted to the new concept by retraining it on the window where the drift was detected. Here, the imbalance module is also invoked to balance the window if necessary. The training happens again from scratch as a result the classifier forgets the previously learnt concept. If the drift is not detected, the classifier continues to learn incrementally on the new window making it suitable for the current concept. The model then reads a new window and the process continues till the stream is exhausted.

## 5.2 EXPERIMENTAL SETUP FOR CDACI

The different parameters required for the working of CDACI are the size of the window for different datasets and also the threshold which is helped to detect the occurrence of the drift. The datasets used for evaluation of model are explained in Section 5.2.1. The experimentation performed to decide window size and threshold is discussed in Section 5.2.2 below.

### 5.2.1 Datasets for CDACI

CDACI is evaluated on real as well as synthetic datasets. The datasets are imbalanced and contains drifts (refer Section 7.2 for detailed descriptions on dataset). The datasets are briefly described with their properties in the Table 5-1 below:

Table 5-1: Dataset properties for CDACI

| Data sets | No. of instances | Imbalance ratio | Type of drift |
|---|---|---|---|
| SEA1 [52] | 100000 | 1.79 | Sudden |
| Electricity [99] | 45312 | 1.36 | Implicit |
| Hyperplane 1 [98] | 100000 | 1.01 | Incremental |
| Hyperplane 2 [98] | 100000 | 1.01 | Subtle Incremental |

The different types of drifts are also mentioned along with number of instances, number of attributes and imbalance ratio. For electricity dataset the exact type of drift is not known hence the type mentioned is implicit

### 5.2.2 Experimentation for threshold and window size

For each dataset, the threshold value for detecting the drift and the window size to read a data stream are decided after experimenting with various options. The various thresholds and the window sizes that are considered are shown in the Figure 5-3 below:

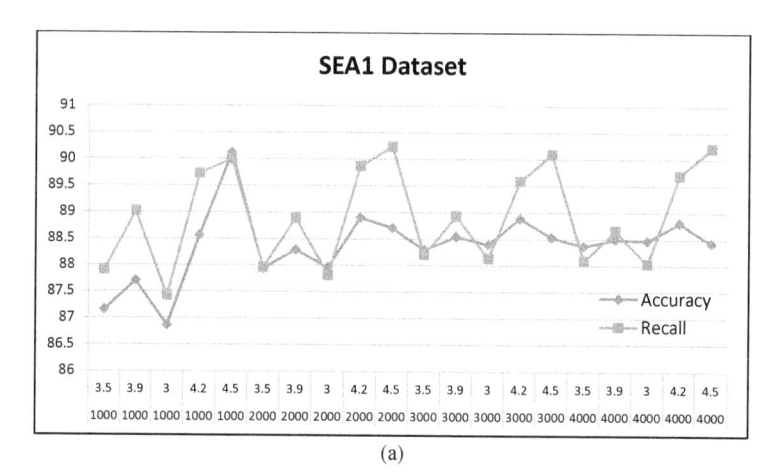

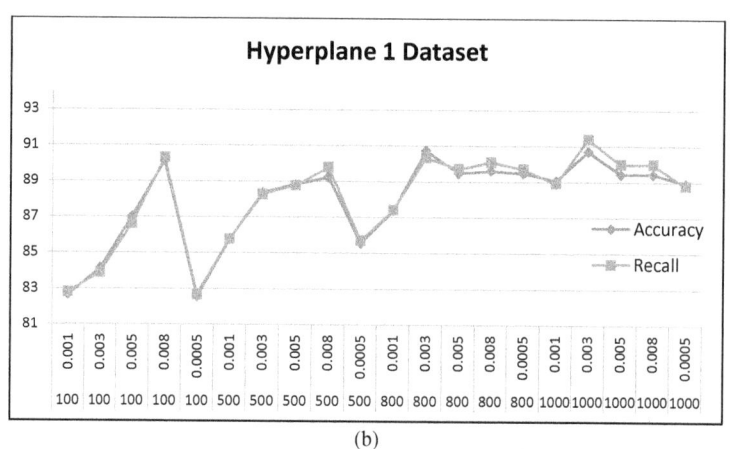

(a)

(b)

## Hyperplane 2 Dataset

(c)

## Electricity Dataset

(d)

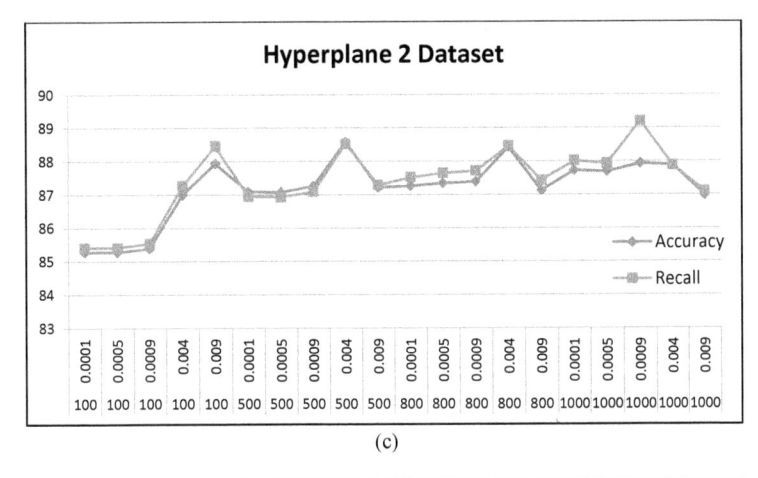

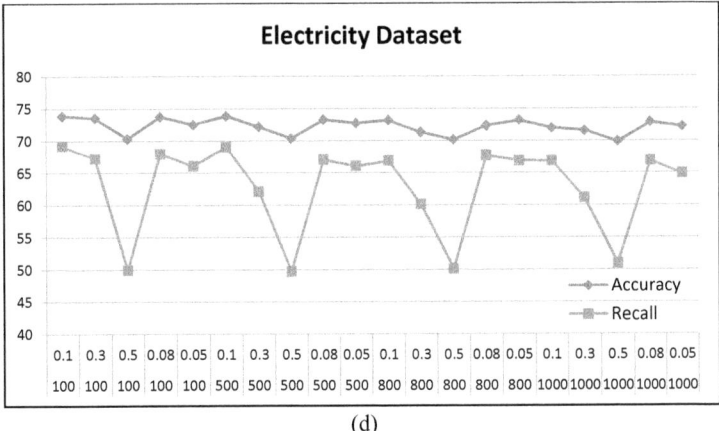

**Figure 5-3: Accuracy and Recall values for different threshold and window size**

For different window sizes i.e. number of instances per window, different threshold values are compared. Threshold values are selected w.r.t to the type of drifts. For sudden drifts larger threshold values ranging from 3 to 4.5 are considered, whereas for incremental and subtle incremental very small threshold values ranging from 0.0005 to 0.009 are considered.

The values that give good recall and accuracy are considered to be optimum values for window size and threshold. After comparing the accuracy and recall values for the different preset values of window size and threshold, the optimum values for the window size and threshold for the respective dataset are as shown in Table 5-2.

Table 5-2: Optimum values of window size and threshold for CDACI

| Dataset | Threshold | Window |
|---|---|---|
| SEA1 | 4.5 | 1000 |
| Electricity | 0.08 | 1000 |
| Hyperplane 1 | 0.003 | 1000 |
| Hyperplane 2 | 0.004 | 1000 |

## 5.3 RESULTS AND CONCLUSION OF CDACI

The performance of CDACI was evaluated for accuracy and recall of the minority class. The results were compared with Gaussian Naïve Bayes (GNB) and Logistic Regression (LR) classifier which are one of the state-of-the-art classifiers. The comparison values of accuracy and recall are as shown in Table 5-3 below:

Table 5-3: Final Accuracy and Recall for CDACI

| Dataset | Classifier | Accuracy | Recall |
|---|---|---|---|
| Electricity | LR | **74.84** | 57.88 |
|  | GNB | 73.39 | 43.25 |
|  | CDACI | 73.84 | **69.19** |
| SEA1 | LR | 88.74 | 92.74 |
|  | GNB | 88.42 | **94.35** |
|  | CDACI | **88.78** | 90.64 |
| Hyperplane 1 | LR | 89.4 | 91 |
|  | GNB | 89.9 | 90.13 |
|  | CDACI | **90.71** | **91.39** |
| Hyperplane 2 | LR | 85.2 | 83.2 |
|  | GNB | 83.5 | 83.36 |
|  | CDACI | **87.89** | **87.88** |

CDACI is able to improve the performance in terms of accuracy as well as minority class recall for Hyperplane1 and Hyperplane2 which has incremental and subtle incremental drifts as compared to LR and GNB. For Electricity dataset, the recall for minority class is increased considerably; however the overall accuracy is better for LR. For SEA1 dataset, the accuracy of CDACI is at par with other classifiers however, the recall for GNB was better.

As compared to HECMI, CDACI did improve the accuracy and recall for minority class considerably due to its drift detection and adaptation techniques. There are still certain limitations of CDACI which are described below:

## 5.4 LIMITATIONS OF CDACI

Though CDACI was successful in detecting and adapting to concept drifts in data streams, it had certain limitations.

1. **Dependency on classifier accuracy:** The drift detection in CDACI totally depends upon the accuracy of the classifier. The base classifier is an in-built Logistic Regression classifier and is not modified for streams. There might a greater scope in increasing the accuracy as well as drift detection by making changes to the classifier.

2. **Dependency on threshold values:** The drift detection in CDACI also depends on the threshold value which is predefined and set experimentally. Detection techniques independent of such preset threshold values can be helpful in improving the performance of any model.

3. **Dependency on window size:** The window size for each dataset is predefined using experimentation. For real time data stream applications such experimentation is not possible. Some heuristics to create dynamic windows can be used.

4. **Assumes fully labeled data:** CDACI assumes that the data stream is fully labeled at all times. Hence it performs supervised classification. However, the data streams face a problem of unavailability of labels.

## 5.5 SUMMARY

CDACI is an incremental model that can detect and adapt to drifts in the stream. It reads the streams using window based technique. Like HECMI, CDACI also assumes that the data is fully labeled. Hence both are supervised classification models. CDACI improves the performance as compared to HECMI considerably by adapting to the drifts. However, it depends on the performance of the in-built classifier and also on the threshold for respective window size. A learning model independent of these predefined thresholds is required. The model should also be able to learn in the absence of fully labeled data.

In the next chapter, a semi-supervised model is proposed to classify imbalanced data streams. This model overcomes the limitations of both HECMI as well as CDACI. It not only detects sudden, gradual and incremental drifts, but also performs learning and classification in the absence fully labeled data.

# Chapter 6

Semi-Supervised Clustering based Classifier
Model - ASSCCMI

# 6. SEMI-SUPERVISED CLUSTERING BASED CLASSIFIER MODEL - ASSCCMI

The models discussed earlier, viz., HECMI and CDACI, were able to deal with the individual problems of imbalance and imbalance with drift handling respectively. The design approaches and the limitations in each one of them aided in taking some major design decisions for this model. Both, HECMI and CDACI, assume fully labeled data and so they are supervised in nature. One of the objectives of the research is to classify in the presence of scarcely labeled data. A limited number of data is assumed to be labeled and a large number of data is unlabeled. In this situation, a semi-supervised approach provides best results. The need of semi-supervised learning is justified further in the next section.

## 6.1 NEED FOR SEMI-SUPERVISED LEARNING

Learning in data streams is possible using supervised or unsupervised learning techniques. In the absence of fully labeled data, supervised learning algorithms can be trained on the limited labeled data and used for classification. When the missing label or unlabeled data is very less as compared to the labeled data, supervised learning algorithms are able to learn and perform classification with good accuracy. However, the problem arises when the percent of labeled data is very less. In the absence of enough labeled data if these supervised learning algorithms are trained, they then face the problems of overfitting. Accuracy of the algorithm drops considerably on unseen data.

In such cases, unsupervised learning algorithms which do not require any class labels can be used. These algorithms are able to learn from labeled or unlabeled data. An efficient unsupervised learning algorithm can create good quality non-overlapping components or cluster by grouping similar data together. These groups can be regarded as classes. But to perform classification an external expert (possibly human) is required to identify and label these groups.

One can also try to obtain the labels for unlabeled data. But, there is a large number of unlabeled data and obtaining labels is expensive in terms of employing experts to label the data and also in terms of the time and effort required. Hence for data streams, learning algorithms that can use labeled and unlabeled data are most suitable. Semi-Supervised Learning (SSL) approaches facilitate such learning from labeled as well as unlabeled data [69]. These approaches can use the scarcely available labeled data and also exploit the advantages of unlabeled data. Semi-supervised learning approaches can deliver high performance of classification by utilizing unlabeled data [103]. Typically, SSL improves classification by using the unlabeled data points and improves clustering by using labeled data points belonging to a cluster [104].

Considering that the research problem is to classify imbalanced data streams with unavailability of fully labeled data, this research recommends the use of semi-supervised learning to handle the problem of imbalanced data streams with scarcely labeled data and proposes an Adaptive Semi-Supervised Clustering based Classification Model to classify Imbalanced data streams (ASSCCMI).

## 6.2 DESIGN DECISIONS FOR SEMI-SUPERVISED LEARNING MODEL

The SSL techniques involve different approaches as discussed in Section 2.8. After analysing these approaches, the cluster and label approach is found to be more applicable for the research problem. The choice of cluster and label approach was also justified through the following reasons:

1. Cluster and label allows exploiting the advantages of having a large number of unlabeled data and few labeled samples. Clustering process helps to create clusters of similar data and labeling process then labels the unlabeled data. This approach applies the cluster assumption of semi-supervised learning and labels the unlabeled data.

2. It is possible to identify a cluster for each class label. Even with a very low percent of labeled data clusters can be created for each class label.

3. Cluster and label approach is more suitable in handling the different possibilities of data appearing in a batch of data stream, as analyzed in section 2.3. Irrespective of the distribution of the class, cluster and label can be used to create clusters for the majority and minority class and then propagate the label to the unlabeled data.

4. Expectation Maximization which optimizes a clustering technique can be used to cluster the data in cluster and label approach. The cluster assumption can be generalized on the training data such that the decision boundary can be found in the low density region to create distinct non-overlapping clusters.

In the following sub-sections design decisions for the proposed semi-supervised learning model, ASSCCMI are explained and justified.

### 6.2.1 Chunk-based Ensemble approach for ASSCCMI

Ensemble model has been found to give better and more accurate results than a single classifier (refer Table 2-1) for many different applications. Moreover, the ensemble model is advantageous when considering imbalanced data as is evident from the HECMI model previously discussed. An ensemble of the clustering model is developed to create clusters. Each model in the ensemble is trained on a batch of data (chunk) stream read from a source. Since the data is partially labeled, collecting the data first into batches/chunks and then training a model on the batch proves to be more appropriate than an online learning. It was also observed from the performance of CDACI that batch based learning is useful in detecting drifts in the stream. The size of each batch is set to fit with available memory. After training and creating clusters, the model only stores a cluster summary and discards the batch of data, thus satisfying the requirements of the memory constraints of data streams.

### 6.2.2 Cluster and Label approach for ASSCCMI

Traditional cluster and label approach includes first clustering the data and then classifying using labeled data [105]. The labeled and unlabeled instances are first formed into clusters using any suitable clustering method. A classifier is then trained on the labeled instances of the cluster. This classifier is then used to predict the labels of the unlabeled instances. Finally, a new classifier is trained on the fully labeled data. This approach is useful when the labeled data contains instances of all classes present in the batch. However, this is not always the case in real time data streams as is observed in Section 2.3. It has been seen in the experimentation that the labeled data might not contain labeled instances of all the classes that have appeared in the unlabeled data. And vice versa, the unlabeled instances might not belong to the class present in labeled data.

Another approach is to cluster data using a suitable clustering technique and then label each cluster using the majority of labeled points in the cluster [106]. The label of the cluster is the label of unlabeled data. However, in case of clusters with no labeled data a suitable label propagation technique should be used.

Almost all different implementations of cluster and label approach in the literature suggest using a suitable clustering technique to form clusters and then labeling the unlabeled instances through a trained classifier or through appropriate label propagation technique. In the proposed ASSCCMI, a novel cluster and label approach is designed to deal with the problem for data streams. Clustering is performed using Expectation Maximization technique. The semi-supervised learning based EM (SSL_EM) is applied on labeled and unlabeled data. The SSL_EM applied helps to create clusters by optimizing the objective function [107] [108]. However, unlike traditional cluster and label approach, the label propagation does not happen for the clusters trained on the same batch. Rather the label is propagated after training three models on 3 consecutive batches of data that consists of labeled and unlabeled data. Labeled and unlabeled clusters from last 3 contiguous models are collected and the labels are propagated from labeled to unlabeled clusters. This is primarily done due to the observations that

1. The labeled data might contain instances of only one class whereas the unlabeled may belong to different classes.

2. The labeled data might contain instances of all the classes whereas the unlabeled data may belong to only one class.

Apart from these, the possibilities of different distribution of classes and no requirement of prior knowledge of ratio of distribution makes cluster and label the obvious choice of semi-supervised learning approach.

A semi-supervised clustering based classifier model that employs an ensemble of cluster and label approach which performs semi-supervised clustering using SSL based Expectation Maximization on a batch of data stream is recommended. The model is adaptive in nature that adapts to the changes in the concept after detecting them.

# Semi-Supervised Clustering based Classifier Model - ASSCCMI

The outline of the remaining chapter includes describing in detail each module of the ASSCCMI model in the order as shown in Figure 6-1.

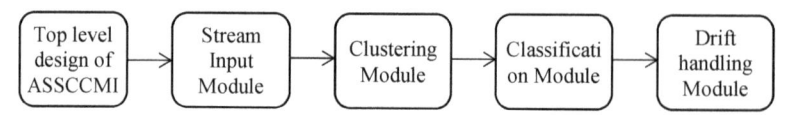

**Figure 6-1: ASSCCMI Chapter Organization**

Flow: Top level design of ASSCCMI → Stream Input Module → Clustering Module → Classification Module → Drift handling Module

## 6.3 TOP LEVEL DESIGN OF ASSCCMI

ASSCCMI - Adaptive Semi-Supervised Clustering based Classification Model for Imbalanced data streams (ASSCCMI) is a semi-supervised framework proposed to solve the problem of classifying imbalanced data streams. The top level design depicting various modules in the framework is as shown in Figure 6-2.

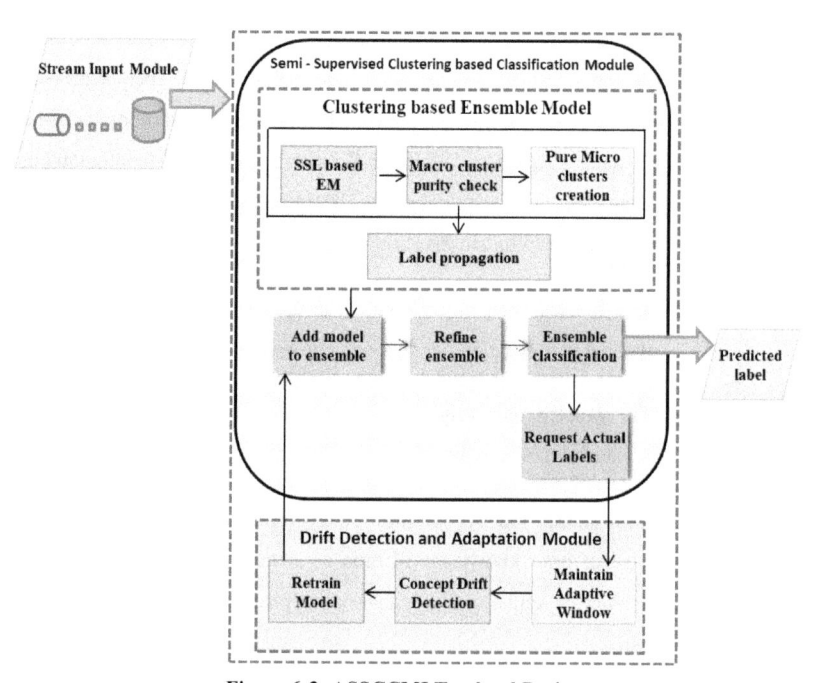

**Figure 6-2: ASSCCMI Top level Design**

The ASSCCMI is an ensemble of classifiers that are trained using the semi-supervised learning technique. Each model in the ensemble is a clustering based model that creates a specified number of clusters for the data seen. The SSL based Expectation Maximization (SSL_EM) helps to create clusters by optimizing the objective function. These clusters are called macro clusters. Once the clusters are created they are checked for purity using the degree of cohesiveness. The macro clusters are then split into homogeneous (pure) micro clusters if found to be impure. Each micro cluster is labeled with the maximum class label found in the cluster. If a cluster contains more number of unlabeled instances then a temporary label of -1 is given to the micro_cluster. The summary of each of the micro clusters is maintained by keeping information about the cluster like {centroid of each dimension, label, total number of instances}. The raw data points are then discarded thus helping to conform to the memory restrictions of data stream mining. Label propagation is performed from labeled micro clusters to unlabeled micro clusters. The ensemble model essentially is then a collection of these micro clusters and classification of unseen data is performed using these micro clusters. The label of the micro cluster closest to the unseen data is given as the final classification by the model.

Once classified, labels of a few selected instances for whom the ensemble model has low confidence are requested. Every batch is assumed to be partially labeled; hence it also receives actual labels for some of the labeled data. Using these actual labels and predicted labels, the model performs concept drift detection. An adaptive window is used to maintain instances of recent windows. The size of the window shrinks when a drift is detected, discarding the instances it has collected and the size of the window grows, when a drift is not detected. Drift detection depends upon the difference in the mean square error of the current window with the average mean square errors of the past windows. On detection of the drift, a clustering model is retrained on the window in which the drift was detected. The retrained model is added to the ensemble, thus making it adapted to the change in the concept. If drift is not detected the model is retrained on the misclassified instances from the classified batch. Thus making it conform to the anytime prediction requirement of data stream mining. The ensemble is then refined before further classification.

The ASSCCMI overall algorithm is shown in Table 6-1.

The data is read into batches through Stream Input Module **Stream_input()**. This module synchronizes the data between the stream source and clustering module. It also ensures that the clustering module always reads the current data by overwriting the older data.

# Semi-Supervised Clustering based Classifier Model - ASSCCMI

**Table 6-1: ASSCCMI Overall Algorithm**

| Algorithm 1: ASCCMI |
|---|
| **Input:** |
|     1. $Z_t$: A partially labeled batch of size n read at time t |
|     2. M: Ensemble size i.e. number of models to be trained in the ensemble |
| **Output:** |
|     1. Ensemble model $C'$ |
|     2. Predicted label $\tilde{Y}$ |
| **Steps:** |
|     1. For i = 1 to M |
|         a. $Z_t \leftarrow Stream\_input()$ |
|         b. $MC_i \leftarrow SSL\_EM(Z_t)$     //Model Ci created using SSL_EM |
|         c. $MiC_i \leftarrow Purity\_check(MC_i)$ //Split clusters into pure clusters |
|     2. Label_prop (MiC) |
|     3. C ← Add_to_Ensemble (MiC$_i$, i) |
|     4. Read Next batch $Z'_t \leftarrow Stream\_input()$ |
|     5. For all $\tilde{X} \in Z'_t = (X'_t, Y'_t, \tilde{X}'_t)$ |
|         a. $\tilde{Y} \leftarrow Classify(C, Z'_t)$ |
|         b. $\tilde{Y}'_t \leftarrow Request\_label()$ |
|         c. $\tilde{Y} \leftarrow \tilde{Y} \cup \tilde{Y}'_t$ |
|     6. $Adaptive\_Window(W, Z'_t, \tilde{Y})$ |
|     7. $Detect\_drift(W, Z'_t, \tilde{Y})$     //Drift detection |
|     8. $Retrain\_ensemble(C, Z'_t, \tilde{Y})$     //Drift Adaptation or Update Ensemble |
|     9. $C' \leftarrow Refine\_ensemble(C, Z'_t, \hat{Y})$     //Refine Ensemble |
|     10. Return $C'$ |

A semi-supervised clustering based classifier MC$_i$ is trained on the partially labeled batch of data read from the input buffer using **SSL_EM()**. Each model is checked for purity to create pure micro cluster before label propagation in **Purity_check()**. Label propagation is performed after training 3 models through **Label_prop()**. The model is then added to the ensemble of classifier **Add_to_Ensemble()**. Once the ensemble is ready, a new batch is read and classification is performed **Classify()**. The new batch is also stored in an adaptive window **Adaptive_Window()**. Every unseen batch after classification is added to the adaptive window. Drift is detected on current batch **Detect_drift()**. If a drift is detected, a new model is trained on the window in which drift is detected, else the model is retrained on misclassified instances using **Retrain_ensemble()**. The ensemble is then refined to maintain the total number of clusters in the model by **Refine_ensemble()**.

ASSCCMI primarily has 3 modules viz., **Stream Input Module, Semi-Supervised Clustering based Classification module and Drift Detection and Adaptation**

**module**. The Semi-Supervised clustering based classification module (SSC) consists of essentially two parts viz., **Clustering based ensemble model** and a **classification module**. The clustering based model is an ensemble of clustering models created by using the SSL based EM and the classification module uses this ensemble to perform classification as well as update and refine the ensemble.

Each of the modules in the framework is described in detail in the following section.

## 6.4 STREAM INPUT MODULE

Stream input module reads data from a stream generator at constant time interval **t** and stores it in a data buffer on a first come first basis, which is then given to the model for training/testing. The stream input module executes two tasks in parallel viz., continuous reading of data from a Stream Generator and executing the ASSCCMI model. The Stream Generator continuously generates the data which is being read and stored in the buffer. The size of the buffer is twice the size of the batch. Initially, the buffer is empty and the stream input module starts storing the data as it is read. There are three possibilities in this situation.

    a. Buffer is not full, reading data from a stream and model is ready.

    b. Model is ready and reading from buffer.

    c. Buffer is full but model is busy.

If the buffer is not full and the model is ready, the input module fills the buffer and meanwhile the model waits for the buffer to get full as shown in Figure 6-3 (a). Once the data buffer is full, the model reads the data from the buffer equal to its batch size clearing this data from the buffer as shown in Figure 6-3 (b). The remaining data moves up in the buffer and new data is appended. Now, if the buffer is full and model is not ready, then the input module overwrites the oldest data in the buffer. While overwriting, the data in the buffer shifts up, removing the oldest data. Shown in Figure 6-3 (c), the selected data shown in red highlight is removed and the new data is added at the end of the list. This helps to remove the oldest data and keep only the current data in the buffer thus enabling the input model to be always apt for the current stream.

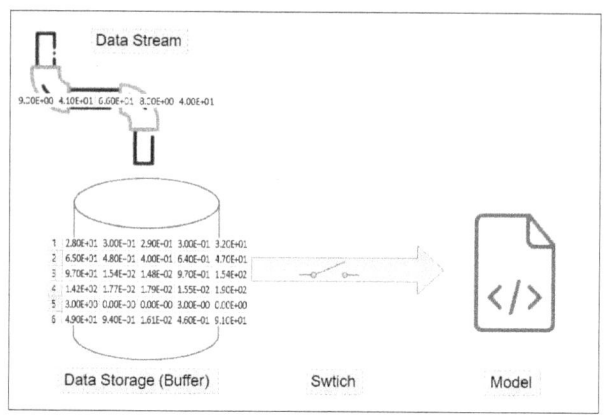

(a) Reading data from a stream in buffer

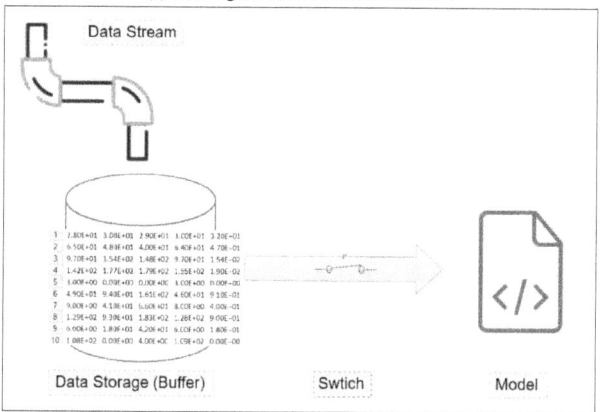

(b) Model ready and reading data from buffer

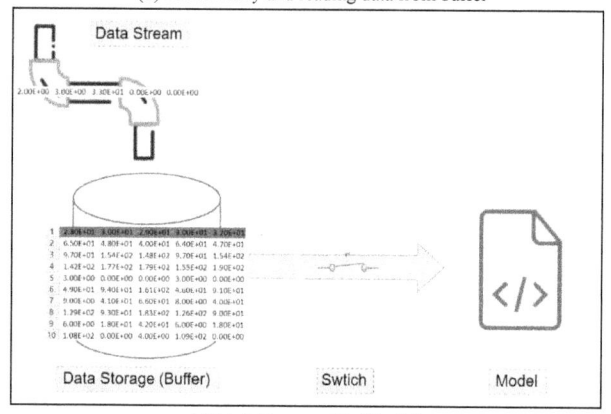

(c) Model busy and buffer full

**Figure 6-3: Stream Input Module**

The Stream Input algorithm is shown in Table 6-2 below.

**Table 6-2: Stream Input Algorithm**

| Algorithm 2: Stream_input() |
| --- |
| **Input:** |
|     Instances of data $\{(x_1, y_1)\}$ |
| **Output:** |
|     Stream $Z_t$ |
| **Steps:** |
|   1. Read instances at time t append into a buffer of size n |
|        $Z_t \leftarrow \text{append}((x_t, y_t))$ |
|   2. If Model C is ready then, |
|   3. Case 1: If buffer full then read $Z_t$ |
|   4. Case 2: If buffer not full then wait for buffer full |
|   5. If buffer ready then, |
|   6. Case 1: If model ready then, SSL_EM($Z_t$) |
|   7. Case 2: If model not ready then, overwrite oldest buffer data |

## 6.5 CLUSTERING BASED ENSEMBLE MODEL

An ensemble of **m** models {**C1, C2 ... Cm**} is created where each model **Ci** is trained using semi-supervised learning based Expectation Maximization technique. Each model **Ci** is trained on a partially labeled batch of size **n**. The batch input $Z_t$, given to each model is defined formally as:

$$Z_t = \{(X_t, Y_t) \cup \widetilde{X}_t\} \quad (6.1)$$

Where,

$X_t = \{x_1, x_2, x_3, \ldots, x_l\}$ are the labeled instances of size **l** and

$Y_t = \{y_1, y_2, y_3, \ldots, y_l\}$ are their respective labels at time **t**

$\widetilde{X}_t = \{x_{l+1}, x_{l+2}, x_{l+3}, \ldots, x_n\}$ are the unlabeled instances of size (**n – l +1**)

The initial **m** models of the ensemble are built from the first **N** data batches. The steps 1-3 in the Algorithm 1 above describe the steps to build the ensemble. For each batch $Z_t$ read from the stream input module, SSL_EM() performs the clustering creating k macro clusters. Each macro cluster is then checked for its purity, to verify if the cluster is cohesive i.e. consisting of instances belonging to the same class or not. If not, the macro clusters are split into pure micro clusters. Label propagation is then performed between the labeled and unlabeled micro clusters. The model is shown in Figure 6-4.

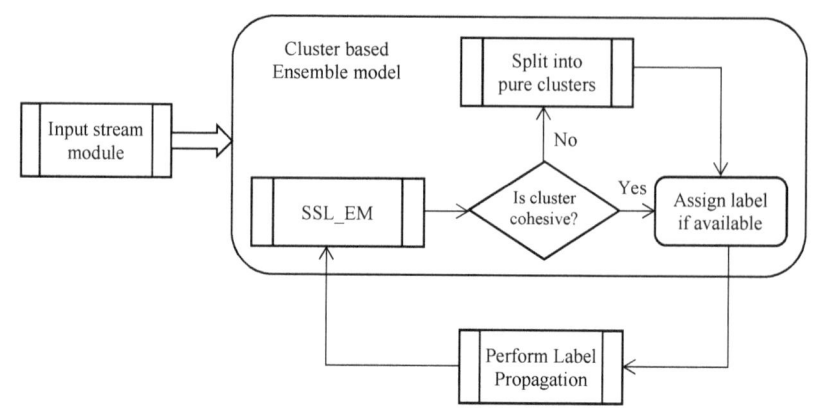

**Figure 6-4: Cluster-based Ensemble model**

### 6.5.1 SSL_EM clustering technique

Training of the model occurs by applying semi-supervised cluster and label approach. The clustering is applied on each batch read. EM extended for semi-supervised learning is used for clustering the data.

Considering the absence of fully labeled data, the data input is defined in (equation 6.1) earlier. Now, for unlabeled data $\tilde{X}_t$, its corresponding label $\tilde{Y}_t$ is unavailable. In case of missing values i.e. Missing At Random (MAR) values or values with some amount of latency, EM proves to be the best suited technique. This is because EM allows probabilistic or soft assignments to these MAR values [107]. In case of data streams, when the data is not fully labeled EM is seen to be efficient at calculating the joint probabilities. Thus Expectation Maximization **becomes the obvious choice** for clustering technique.

The Expectation Maximization method traditionally consists of three steps viz., the initialization step, the expectation step and the maximization step. The initialization step consists of initializing the number of clusters and the centroids of each cluster. An improper initialization can have an adverse effect on the EM algorithm resulting into poor clusters. The expectation step (E-step) estimates the missing value/expected value of latent variable. The maximization step (M-step) optimizes the parameters of probability distribution. The E and M step iterate over till convergence criteria is met.

In the SSL_EM, the objective of EM is to optimize the assignment of labeled and unlabeled instances to the clusters. The steps in SSL_EM are similar to the traditional EM steps. The probability of the unknown variable $\tilde{Y}_t$ given known variables $X_t, Y_t$ and $\tilde{X}_t$ is estimated. The initialization, expectation and maximization steps are discussed below. The E-step and M-step are iteratively performed till a convergence criterion is met. The convergence criterion is generally when there is no change in the parameters and a maximum likelihood is achieved.

### *6.5.1.1 Initialization*

Instead of the usual process of initializing the clusters randomly, the availability of labeled instances in the data is utilized for added advantage. The number of clusters is initialized to K. After experimenting with different number of clusters an optimum value of K is selected for each dataset. For every class in the data buffer, the initial number of clusters assigned to that class is proportional to the ratio of the number of labeled instances of that class and the total number of labeled instances in the batch. If $N_c$ is the number of instances with label **c** and N is the total number of labeled instances in the batch, then the number of clusters assigned to class **c** are calculated as: $k_c = K (N_c / N)$. Considering that the labeled data can be as low as 5% or as high as 90% of the total data instances, the initial number of clusters can be less than the labeled instances of a particular class i.e. $k_c < N_c$ or the initial number of clusters can be more than the labeled instances of that class i.e. $k_c > N_c$. If the number of available labeled instances of a class, $N_c$, is more than the number of clusters i.e. $k_c < N_c$, then the $k_c$ clusters are initialized by randomly selecting instances from $N_c$ as centroids. And if the number of labeled instances of a class, $N_c$, is less than the total number of clusters i.e. $k_c > N_c$, then the $k_c$ clusters are initialized using the $N_c$ instances and remaining ($k_c - N_c$) clusters are initialized by selecting randomly from the unlabeled instances. The initial centroids, $\hat{\mu}_c$, the respective label **c** and data instances for each cluster is stored.

If the number of clusters is divided equally among the class label, then the minority class creates clusters which overlaps with majority class cluster, whereas the majority class creates sparse and dense clusters. Hence, this initialization technique is recommended for ASSCCMI and is found to optimize the clusters.

## 6.5.1.2 Expectation Step

In this step, the EM assigns instances to the cluster initialized in the earlier step. The EM is expected to optimize this assignment. Since there is labeled and unlabeled data, the assignment of data to clusters is different than traditional EM. Here, the fact that data is labeled is used for the benefit of creating clusters for each class. First, the formula for assigning elements to the cluster is derived and then using the formula assigning labeled and unlabeled data to clusters is explained.

**Deriving the EM formula:**

The data in a data streams generally is not considered to be IID (Independent and Identical Distribution) due to the voluminous nature and due to the fact that they are generated from different sources. However, the stream is read in batches and the size of each batch is limited, hence it is safe to assume that the samples drawn from a batch are all IID and are generated from the same distribution P(x). And hence,

For any $x_i \in X$,

$$P(x_0, x_1, x_2, \dots, x_i) = \prod_{0}^{i} P(x_i) \qquad (6.2)$$

And the joint probability of features $x_i$ and label $y_i$ is given by

$$P(x,y) = P(x).P(y) \qquad (6.3)$$

Now, the complete-data likelihood when the $\tilde{y}_i$ are known is given as

$$P(X, Y, \tilde{X}, \tilde{Y}) = \prod_{i=1}^{l} p(x_i, y_i) \times \prod_{i=l+1}^{n} p(\tilde{x}_i, \tilde{y}_i) \qquad (6.4)$$

But since the values of $\tilde{y}_i$ are not known, EM is used to compute the probabilities of the unknown variables and to optimize the assignment of instances to the clusters.

Now, given the initial centroids for each cluster $\hat{\mu}_c$, the labeled and unlabeled instances, the E-step is used to find the posterior distribution of the unknown variable. This is done by computing the expected complete data log-likelihood $Q(\mu, \hat{\mu}_c)$ as:

$$Q(\mu, \hat{\mu}_c) = \mathbb{E}_{\tilde{Y}|X,Y,\tilde{X},\hat{\mu}_c}[\log p(X, Y, \tilde{X}, \tilde{Y} | \mu)] \quad (6.5)$$

Where,

$\hat{\mu}_c$ — initial centroids of cluster with label c

$\mu$ — centroid that EM intends to optimize

X, Y — data instances and their respective labels

$\tilde{X}, \tilde{Y}$ — unlabeled instances and unknown labels

Continuing with the equation (6.5), the expected log likelihood can be written as

$$Q(\mu, \hat{\mu}_c) = \sum_{\tilde{Y}} p(\tilde{Y}|X, Y, \tilde{X}, \hat{\mu}_c) \log p(X, Y, \tilde{X}, \tilde{Y}|\mu)$$

$$Q(\mu, \hat{\mu}_c) = \sum_{\tilde{Y}} p(\tilde{Y}|X, Y, \tilde{X}, \hat{\mu}_c) \left[ \sum_{i=1}^{l} \log p(y_i, x_i|\mu) + \sum_{i=l+1}^{n} \log(p(\tilde{y}_i, \tilde{x}_i|\mu)) \right]$$

$$Q(\mu, \hat{\mu}_c) = \sum_{i=1}^{l} \log p(y_i, x_i|\mu) \sum_{\tilde{Y}} p(\tilde{Y}|X, Y, \tilde{X}, \hat{\mu}_c)$$

$$+ \sum_{i=l+1}^{n} \sum_{\tilde{Y}} p(\tilde{Y}|X, Y, \tilde{X}, \hat{\mu}_c) \log p(\tilde{y}_i, \tilde{x}_i|\mu) \quad (6.6)$$

For a given centroid $\hat{\mu}_c$, unlabeled data $\tilde{X}$ and labeled data X, Y; it is known that the sum of the conditional probabilities for $\tilde{Y}$ is equal to 1. Hence, in the first term of equation (6.6),

$$\sum_{\tilde{Y}} p(\tilde{Y}|X, Y, \tilde{X}, \hat{\mu}_c) = 1$$

Now, the probability that $\tilde{Y}$ is allotted a label c, is completely dependent upon $\tilde{X}$ and $\hat{\mu}_c$. Hence, the second term of equation (6.6) above is given as:

$$\sum_{i=l+1}^{n} \sum_{\tilde{Y}} p(\tilde{Y}|X, Y, \tilde{X}, \hat{\mu}_c) = \sum_{i=l+1}^{n} \sum_{\tilde{y}_i=\{1...c\}} p(\tilde{y}_i|\tilde{x}_i, \hat{\mu}_c)$$

Updating equation (6.6), the complete data log-likelihood becomes:

$$Q(\mu, \hat{\mu}_c) = \sum_{i=1}^{l} \log p(y_i, x_i|\mu) + \sum_{i=l+1}^{n} \sum_{\tilde{Y}} p(\tilde{y}_i|\tilde{x}_i, \hat{\mu}_c) \log p(\tilde{y}_i, \tilde{x}_i|\mu) \quad (6.7)$$

In the above equation (6.7), the first term in the right hand side is the sum of probabilities of all the labeled instances and the second term is the sum of the unlabeled instances. Analysing the equation for labeled and unlabeled instances separately below:

**Labeled part:**

Now, the log likelihood of the labeled instances shown in equation (6.7) can be calculated using the joint probability as:

$$p(y_i = c, x_i | \hat{\mu}_c) = p(x_i | y_i = c, \hat{\mu}_c) \cdot p(y_i = c | \hat{\mu}_c) \qquad (6.8)$$

Where,

$p(x_i | y_i = c, \hat{\mu}_c)$ is the conditional probability

$p(y_i = c | \hat{\mu}_c)$ is the prior probability

For the conditional probability, given the initial centroid for each cluster $\hat{\mu}_c$ with label ($y_i = c$), there are three possibilities of assigning an instance to the cluster:

1. An instance $x_i$ with label $y_i = c$ is assigned to the cluster $\hat{\mu}_c$ with a probability 1, if $x_i$ is closer to $\hat{\mu}_c$ and the label of $\hat{\mu}_c$ is **c** and for all other classes its probability is 0.

2. If the closest cluster does not have label c, then the instance is assigned to the second closest cluster with label c.

3. If none of the clusters have label = c, then the instance is assigned to a closest cluster.

$p(x_i | y_i = c, \hat{\mu}_c)$

$$= \begin{cases} 1; \min(\|\tilde{x}_i - \hat{\mu}_c\|^2) \text{ and Label}(\hat{\mu}_c) = c \text{ or} \\ \quad \text{second closest } (\min(\|\tilde{x}_i - \hat{\mu}_c\|^2)) \text{ and Label}(\hat{\mu}_c) = c \text{ or} \\ \quad \min(\|\tilde{x}_i - \hat{\mu}_c\|^2) \text{ and Label} = \varphi \\ 0; \text{ for all other labels} \end{cases} \qquad (6.8a)$$

The prior probability for the labeled data in the same equation (6.8) is the number of instances belonging to the class $y_i = c$.

$$p(y_i = c | \hat{\mu}_c) = \frac{\text{Count\{No. of instances with label} = c\}}{\text{Count\{Total number of labeled instances\}}} \qquad (6.8b)$$

**Unlabeled part:**

Now for unlabeled instances part of equation (6.7); the probability that an unlabeled instance $\tilde{x}_i$ is assigned to a cluster with centroid $\hat{\mu}_c$ is high, if it is closer to that centroid. The proximity of the instance to the cluster is directly proportional to the similarity between instance and cluster centroid $sim(\tilde{x}_i, \hat{\mu}_c)$. Euclidean distance similarity measure is used to calculate the proximity of the instance $\tilde{x}_i$ with the centroid $\hat{\mu}_c$ of that cluster. Hence,

$$p(\tilde{y}_i|\tilde{x}_i, \hat{\mu}_c) \propto sim(\tilde{x}_i, \hat{\mu}_c)$$

$$p(\tilde{y}_i|\tilde{x}_i, \hat{\mu}_c) = \begin{cases} 1; \ min \ (\|\tilde{x}_i - \hat{\mu}_c\|^2) \\ 0; for \ other \ classes \end{cases} \quad (6.9)$$

The probability of each instance belonging to a particular cluster is thus calculated and instances are assigned to the cluster.

**Assigning the instances to clusters:**

Now the EM assigns the labeled and unlabeled instances to the closest cluster using equation 6.8 a, 6.8 b and 6.9. When all instances are assigned to the clusters, the label of the cluster is reassigned using the *cluster assumption* of semi-supervised learning that states [69] [105]:

*"If two points x1, x2 are in the same cluster they are likely to be in the same class."*

Applying this assumption, the label of maximum labeled instances is assigned to the cluster.

The E-step conclusively thus assigns instances to respective clusters. An account of the labels of maximum labeled instances of each cluster along with the instances belonging to it is maintained in the macro cluster. Once when all the instances are assigned to the cluster, the M-step is invoked.

### 6.5.1.3 Maximization step

The objective of the M-step is to maximize the expectation i.e.

$$\mu_c = \underset{\mu}{\mathrm{argmax}}(Q(\mu, \hat{\mu}_c)) \qquad (6.10)$$

In this step, the cluster centres for each cluster are recomputed by taking the average of instances assigned to it.

$$\mu_c = \frac{1}{N_c} \sum_{x \in X_c} x \qquad (6.11)$$

Where, $N_c$ – number of instances belonging to cluster c.

The EM then continues with the next iteration of E-step with $\hat{\mu}_c = \mu_c$ until convergence.

### 6.5.1.4 Convergence criteria

The EM converges if there is no change in the values of $\mu_c$ for two contiguous iterations or if the difference between the two contiguous centroids is less than a convergence threshold $\varepsilon$ i.e. $\|\mu_c - \mu\| < \varepsilon$. Once the EM converges, the summary of the cluster is stored in a macro cluster in the form of:

$$\begin{aligned}&\text{Macro cluster} \qquad (6.12)\\ &= \{[\mu_1, \mu_2, \ldots, \mu_d], \text{label}, \text{instancesofcluster}, \text{numofinstances}\}\end{aligned}$$

The summary information includes the centroids of all the attributes/dimensions, the label assigned to the cluster considering the label of the maximum labeled instances belonging to the cluster after the EM, the actual instances belonging to the cluster and the total number of instances in that cluster. The label assigned to any cluster is the same as the class label. For clusters containing only unlabeled data, a temporary label of (-1) is assigned.

The SSL_EM algorithm is shown in Table 6-3.

Table 6-3: SSL_EM algorithm

| Algorithm 3: SSL_EM |
| --- |
| **Input:** |
| $Z_t$: A partially labeled batch of data stream of size n read at time t |
| $Z_t = \{(X_t, Y_t) \cup \tilde{X}_t\}$ |
| **Output:** |
| Macro_Cluster MC |
| $MC\{label: c, centroid: \mu_{(c,d)}, Instances\ (X, \tilde{X}), No. of\ elements: N_c\}$ |
| **Steps:** |
| 1. Augment label for unlabeled data |
| $\quad \tilde{X}_t = \{-1, -1, \ldots\ldots. (l + n)\ times\}$  // Label '-1' for unlabeled $\tilde{X}_t$ |
| 2. Find distinct label c from $X_t$ |
| 3. $N \leftarrow |n_i|$ |
| 4. $|n_c| \leftarrow$ Count number of elements of each class |
| 5. Initialization step: |
| 6. Initialize no_of_clusters K = 15   // SSL_EM Initialization |
| $\quad k_c = K\ |n_c|\ /\ |n_i|$ |
| 7. If $k_c < |n_c|$, then $\mu_c \leftarrow$ random (instances of label c) |
| 8. Else If $k_c > |n_c|$, then $\mu_c \leftarrow$ random ($k_c$ instances of label c + remaining random unlabeled instances ) |
| 9. Expectation step: |
| $\quad$ a. Assign instances to the centroids initialized above. |
| $\quad$ b. For labeled instance: |
| $p(x_i|\ y_i c, \hat{\mu}_c)$ |
| $= \begin{cases} 1; \min\ (\|\tilde{x}_i - \hat{\mu}_c\|^2)\ \text{and}\ Label(\hat{\mu}_c) = c\ \text{or} \\ \text{second closest}\ \left(\min\ (\|\tilde{x}_i - \hat{\mu}_c\|^2)\right)\ \text{and}\ Label(\hat{\mu}_c) = c\ \text{or} \\ \min\ (\|\tilde{x}_i - \hat{\mu}_c\|^2)\ \text{and}\ \ \text{Label} = \varphi \\ 0; \text{for all other labels} \end{cases}$ |
| $\quad$ c. For unlabeled instance, assign instances to minimum distance centroid |
| $p(\tilde{y}_i|\tilde{x}_i, \hat{\mu}_c) = \begin{cases} 1; \min\ (\|\tilde{x}_i - \hat{\mu}_c\|^2) \\ 0; for\ other\ classes \end{cases}$ |
| 10. Maximization step: $\mu_c = \frac{1}{N_c} \sum_{x \in X_c} x$ |
| 11. Repeat 7-8, until convergence $\|\mu_c - \mu\| < \varepsilon$ |
| 12. Return |
| $\quad MC\{label: c', centroid: \mu_{(c,d)}, Instances\ of\ clusters, No. of\ elements: N_c\}$ |

### 6.5.2 Macro cluster purity check

The objective of the clustering method is to create pure homogeneous clusters. Clusters containing instances from only one class are pure homogeneous. The higher the purity, the more cohesive are the clusters. For label propagation, it is imperative that the clusters created by SSL_EM are pure. Each macro cluster is checked for high cohesion to ensure high intra-cluster similarity and low inter-cluster similarity using purity coefficient. This ensures correct cluster assignment of instances. A macro cluster is considered to be pure, if it has labels of only one class i.e. either labeled data of one class or only unlabeled data, else it is not pure. Impure clusters results in low quality clustering, suggesting scattered or sparse clusters. As a result of such clusters, the label propagation performs poorly and clusters not similar to one another are assigned same labels. During macro cluster purity check the following conditions are checked and the clusters are split into pure homogeneous clusters if required.

1. If a cluster contains **labels of only one class**, then it is considered to be pure and no splitting is done.

2. If a cluster contains **labels of 2 or more classes**, then the following conditions are checked

    a. The cluster contains **labeled instances of different classes**: Since only labeled instances are present, separate cluster for each respective class label with their corresponding instances can be created.

    b. The cluster **contains labeled and unlabeled instances**: Check the cluster for purity. If the purity coefficient is above a particular threshold, then no splitting is done and the unlabeled instances belong to the labeled cluster; else clusters are split.

    c. The cluster **contains only unlabeled instances**: It might happen that unlabeled instances from two different classes are placed in one cluster. Check the cluster for purity. If the purity coefficient is above a particular threshold, then no splitting is done, else split the cluster.

The cluster purity check function works on the principle of silhouette coefficient. For every cluster, the function calculates the intra_cluster distance between a data element and all other data instances belonging to the cluster. Then it calculates the inter_cluster distance between a data element belonging to a cluster and all elements from other clusters. A minimum average of this distance is calculated. Finally, the purity coefficient is computed by taking the difference between the intra_cluster and inter_cluster distances. The purity coefficient range of values lies between -1 to +1, where -1 is highly impure suggesting none of the elements belong to the cluster they are assigned to and +1 suggests elements have matched their cluster assignment highly. Clusters with purity coefficient > 0.6 are considered to be pure. The detailed algorithm is explained in Algorithm4: Purity_check in Table 6-4 below.

### 6.5.3 Pure Micro cluster creation

Once the macro cluster is checked for purity, depending upon its purity coefficient, it may be split into further pure micro clusters. Apart from the purity coefficient, macro cluster can also be split if it contains labels of 2 or more classes. Splitting of macro cluster is performed as follows:

1. If the purity is high, then the cluster is highly cohesive, it will not be split and the summary information is maintained
2. If the purity is below the threshold, then the labels of the instances are checked.
    a. If cluster contains labeled and unlabeled data, then cluster is split. Separate clusters for labeled instances corresponding to their class labels are created and also another cluster for unlabeled instances is created.
    b. If cluster contains only unlabeled instances, then cluster is split into k-clusters using k-means clustering, where k is equal to the number of classes.

After splitting, each cluster summary is updated to include recalculated centroids, max label, instances belonging to the cluster and total number of instances. Now all clusters are considered to be pure or having high purity coefficient ensuring the clusters are cohesive. These cohesive clusters are now called micro clusters consisting of elements belonging to one class only.

$$\text{Micro cluster} = \{\text{label}, [\mu_1, \mu_2, .., \mu_d], \text{instancesofcluster}, \text{numofinstances}\} \quad (6.13)$$

Each of these micro clusters is a part of the clustering model $C_i$ which is added to the ensemble. The algorithm shown below in Table 6-4 includes the cluster purity check and micro cluster creation steps.

**Table 6-4: Cluster purity check algorithm**

| |
|---|
| **Algorithm 4: Purity_check** |
| **Input:**<br>　　MC: Macro clusters created by SSL_EM()<br>　　$MC\{label: c', centroid: \mu_{(c,d)}, No. of\ elements: N_c\}$<br>**Output:**<br>　　MiC: Labeled and unlabeled pure micro_cluster<br>　　$MiC\{label: c', centroid: \mu_{(c,d)}, No. of\ elements: N_c\}$ |
| **Steps:**<br>1. Measure intra_cluster distance $a_i$, between i and all other points of the cluster to which *i* belongs.<br>$$a_i = \frac{1}{|C_i| - 1} \sum_{j \in C_i, i \neq j} d(i,j)$$<br>2. For all other clusters C, to which i does not belong, calculate the average dissimilarity d(i,C) of i to all observations of C.<br>$$b_i = \min_{k \neq i} \frac{1}{|C_k|} \sum_{j \in C_k} d(i,j)$$<br>3. $P_i = (b_i - a_i) / \max(a_i, b_i)$　　　// Purity coefficient<br>4. $P_c = Avg(P_i)$　　　// Purity for entire cluster:<br>5. If $P_c > 0.6$,<br>　　a. C{label} ← max_label(C)　// Pure cluster<br>　　b. MiC ← C　　　// Add cluster to pure micro cluster<br>6. Else　　　//Cluster is not cohesive<br>　　a. $C_i$ ← split_cluster(C)<br>　　b. $C_i\{label\}$ ← max_label($C_i$)<br>　　c. Ci{centroid} ← updatecentroid($C_i$)<br>　　d. MiC ← C U $C_i$　// Add cluster to pure micro cluster<br>7. Return MiC |

### 6.5.4 Label Propagation

Label propagation is used to label the unlabeled data. The labels are propagated from labeled micro clusters to the neighbouring unlabeled micro clusters.

Labeled micro cluster $\mu_c$ → Label Propagation using Similarity → Unlabeled micro cluster $\hat{\mu}_c$

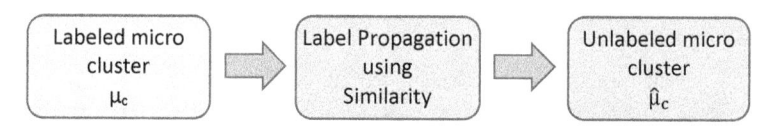

**Figure 6-5: Label Propagation**

The label propagation is based on the *smoothness assumption* of semi supervised learning which states that [69] [105]:

***"If two points x1, x2 in a high density region are close so should be their corresponding outputs y1, y2."***

So, if an unlabeled instance $\tilde{x}$ is closest to a labeled instance $x$ *with label c*, then the unlabeled instance is also labeled as $c$.

Now, the probability that an unlabeled instance $\tilde{x}$ is assigned a label $c$ is 1, if the instance is closest to a cluster with label $c$

Hence,

$$P(\tilde{x} = c|\mu_j) = \begin{cases} 1; c = \text{argmin}_{j=1..c} \|\tilde{x} - \mu_j\|^2 \\ 0; \text{for other clusters} \end{cases} \quad (6.14)$$

Extending this cluster assumption to a whole cluster instead of a single instance, the label of unlabeled cluster can be obtained by:

***"If a cluster is represented by its centroid, then the probability that an unlabeled cluster is assigned a label c is 1, if it is closest to a centroid with label c than any other centroids."***

Hence, the equation (6.13) is updated as:

$$P(\tilde{\mu} = c|\mu_j) = \begin{cases} 1; c = \text{argmin}_{j=1..c} \|\tilde{\mu} - \mu_j\|^2 \\ 0; \text{for other clusters} \end{cases} \quad (6.15)$$

Label propagation is performed between micro clusters of the last m contiguous models. A distance matrix is created between the labeled and unlabeled micro clusters. The labeled micro clusters (LM) form the rows and the unlabeled micro clusters (UM) form the columns of the similarity matrix. The similarity between labeled and unlabeled micro cluster is calculated using Euclidean distance between the centroids. To assign the label to an unlabeled micro cluster, the closest labeled micro cluster is first found by taking the minimum distance between each column j and the corresponding row i. The labeled micro cluster at the row i is identified as the one closest to the unlabeled micro cluster. The label of this $i^{th}$ micro cluster is assigned to the unlabeled micro cluster at j as shown in Figure 6-6 below. After comparing the distance of UM₁ with all labeled micro cluster it is found that the distance between UM₁ and LM₂ is the smallest. Hence, the UM₁ is labeled with the label of LM₂.

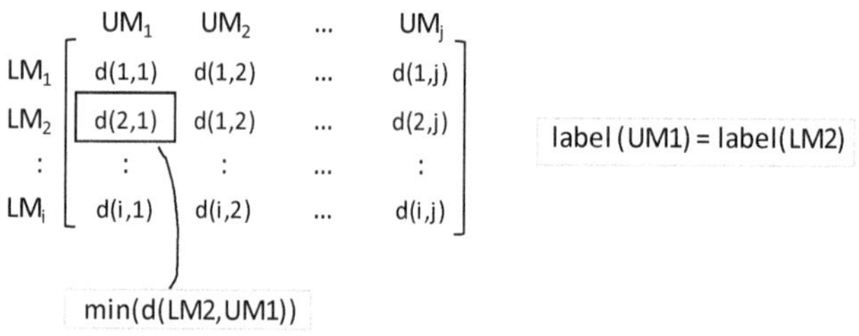

$$\begin{array}{c} & UM_1 & UM_2 & \cdots & UM_j \\ LM_1 & d(1,1) & d(1,2) & \cdots & d(1,j) \\ LM_2 & d(2,1) & d(1,2) & \cdots & d(2,j) \\ \vdots & \vdots & \vdots & \cdots & \vdots \\ LM_i & d(i,1) & d(i,2) & \cdots & d(i,j) \end{array}$$

min(d(LM2,UM1))

label(UM1) = label(LM2)

**Figure 6-6: Label Propagation matrix representation**

Once the labels are propagated, the instances are now discarded and the micro cluster summary of unlabeled data is updated to reflect the new labels.

$$\text{Micro cluster} = \{label, [\mu_1, \mu_2, \ldots, \mu_d], numofinstances\} \quad (6.16)$$

These micro clusters are now added to the ensemble which will be used for classification of unseen data.

Generally label propagation is applied within the batch. But the reason for ASSCCMI using last m models is because of the volatile nature of data streams. Various possible scenarios of data batches with their labeled and unlabeled data instances as well as majority and minority class ratios are analyzed and explained in the section 2.3. There can be scenarios where a) the labeled data might contain instances of only one class whereas the unlabeled data may belong to different classes or b) the labeled data might contain instances of all the classes whereas the unlabeled data may belong to only one class or c) the minority class instances are not seen either in the labeled instances or in the whole batch for the last b batches. If the label is propagated within the batch, there can be incorrect propagation especially when the labeled instances have missing class labels and the batch consists of instances from missing class. Without the prior information about the ratio of distribution of classes, the label propagation within the batch can also have adverse effect. Hence, ASSCCMI recommends propagating the label within last 3 models and not within the same batch on which a model is trained.

The algorithm 5 shown in Table 6-5 explains the label propagation technique of ASSCCMI.

Table 6-5: Label Propagation algorithm

| Algorithm 5: Label_propagation() |
|---|
| **Input:** <br>    MiC: Labeled and unlabeled pure micro_cluster <br>    $MiC\{label: c', centroid: \mu_{(c,d)}, No.of\ elements: N_c\}$ <br> **Output:** <br>    C: Labeled micro_cluster <br>    $C\{label: c, centroid: \mu_{(c,d)}, No.of\ elements: N_c\}$ |
| **Steps:** <br> 1. After every 3 models <br>       LC ← Labeled (MicroC) <br>       C ← LC <br>       UnC ← UnLabeled (MicroC)    // unlabeled micro cluster <br> 2. For each $UnC_i$ in UnC: <br>       $UnC_i\{label\}$ ← closest $LC_j\{label\}$ <br> 3. C = C U $UnC_i$ <br> 4. Return MiC |

## 6.6 CLASSIFICATION

At this stage, the ensemble model is trained and ready for classification. A new batch is first read from the input modules' data buffer and prediction of the classes is performed. First, the closest micro cluster is identified by using the similarity between the unseen sample and all the micro clusters from the ensemble. Then the label of this closest micro cluster is given as the final prediction.

Generally in an ensemble technique, the result of each model in the ensemble is taken and the max prediction from these results is given as the final prediction. However, ASSCCMI doesn't apply this technique and rather it uses the closest micro cluster from the ensemble. This is done keeping in mind the imbalance nature of the data. Due to the presence of large number of majority class sample, each model in the ensemble is bound to have more clusters representing the majority class and very

Semi-Supervised Clustering based Classifier Model - ASSCCMI

few clusters representing the minority class. As a result, most of the models in the ensemble will predict the label of the unseen sample as the majority class label.

Ensemble confidence for each instance in the batch is calculated to identify the accuracy of the ensemble for each of them. All the models in the ensemble predict the class label based on the similarity measure. The confidence of the maximum predicted class label l is measured by taking a percentage of the number of models predicting label l by the total number of models. The actual labels of the instances with confidence measure less than 70% are requested from the expert. The actual labels from the batch together with the requested labels and the predicted labels are then used for drift detection. The Figure 6-7 shown below depicts the steps followed after the ensemble model is created and new batch is read for classification and drift handling.

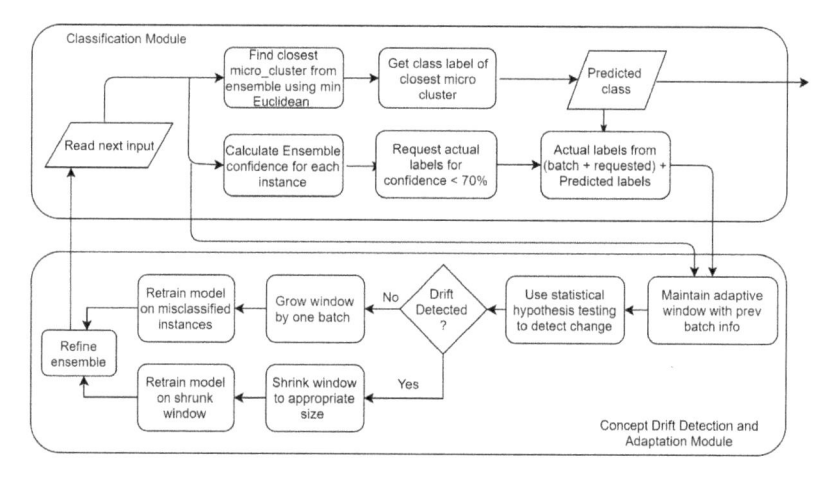

**Figure 6-7: Classification and Concept Drift Handling**

## 6.7 CONCEPT DRIFT DETECTION AND ADAPTATION

Drift detection and adaptation happens after the classification. An adaptive window based method is employed to detect and adapt to drifts. The adaptive window maintains two sections (sub-windows) viz., a fixed section which consists of an aggregate of accuracy and the mean squared error of the previous batch and a sliding section that grows to a max size **win_size** and shrinks. Figure 6-8 depicts the adaptive sliding window designed for ASSCCMI.

## Semi-Supervised Clustering based Classifier Model - ASSCCMI

Every time a new batch is read, it is given to the classification as well as stored in the sliding sub-window. The actual labels and the predicted labels are used to calculate the mean square error. The mean square error on the current batch and previous (m-1) batches are compared using a statistical testing method. If there is no change in the error rates between the past and the current windows, the adaptive window grows, the cumulative accuracy and error are re-calculated and the current batch is added to the existing window. This window grows up to a specific size **win_size**. If the window size reaches to **win_size**, then the oldest batch in the window is pushed appending the current batch and readjusting the cumulative accuracy. If a drift is detected, the window shrinks to a specific size depending upon the type of the drift. The instances from this remaining shrunk window are then used to retrain the model. The new model is then added to the ensemble and the ensemble is refined.

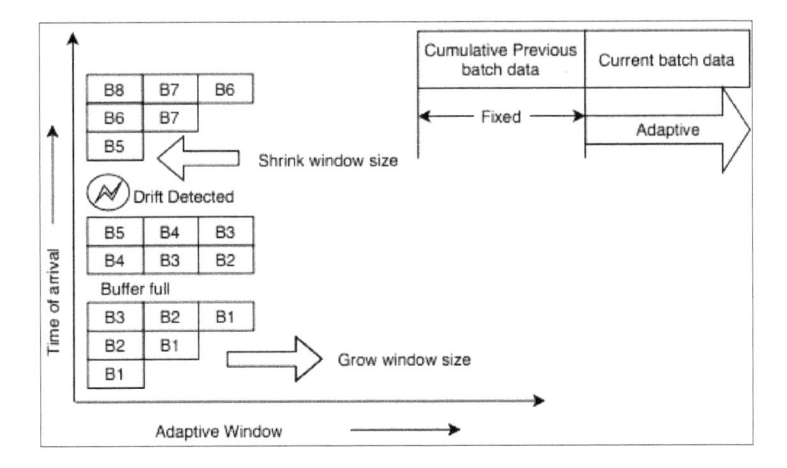

Figure 6-8: Adaptive Sliding window for drift detection

### 6.7.1 Drift Detection

The ensemble model in Section 6.5 is always trained on the recent data stream. This ensures that the model is always updated with the current concept. Hence, it can be safely assumed that, the mean square error rate of the ensemble model is accurately estimated for the previous m-1 batches [109]. A considerable change in the error rate then suggests a change in the concept i.e. the data on which the model was trained earlier has changed. The change can be in the negative direction or in the positive direction. In short, the error rate can be increasing or decreasing. Increase in the error rate implies the ensemble is not able to predict accurately. This change in the error rate can then be used as a parameter for statistical hypothesis testing. This module uses change in the error between two consecutive windows to detect drift in the latest window. Like the CUSUM type of drift detection [13], average error of the data seen so far is maintained in the fixed section of the adaptive window. The drift detection is performed using statistical hypothesis testing between two consecutive windows. The null hypothesis H0 and the alternate hypothesis H1 for this statistical test are as follows:

H0 → there is no change in the validation error rate of the two windows

H1 → there is a change in the validation error rate of the two windows

If the validation error rate changes, the null hypothesis is rejected and concept drift is said to have occurred. Otherwise, no drift is said to have occurred.

The error rate of the model till time t-1 is given by E(t-1). It includes the cumulative mean of the error rate which the model calculates for the data till (t-1). The error rate of the model on the current window is given by E(t). The number of instances for previous window are given by N(t-1) and the number of instances for the current window N(t).

Let $\hat{d}$ be the difference in the error rate between the previous window and the current window. The $\hat{d}$ is then given by

$$\hat{d} = E(t) - E(t-1) \qquad (6.17)$$

This difference $\hat{d}$ in the error rate gives an unbiased estimate of the mean **d** [110].

The variance of this distribution is the sum of the variances of E(t-1) and E(t). The variance is then given by

$$Var\,(\sigma^2) = \frac{E(t)[1-E(t)]}{|N_t|} + \frac{E(t-1)[1-E(t-1)]}{|N_{t-1}|} \qquad (6.18)$$

For any data instance, to obey the normal distribution with variance $\sigma^2$, standard deviation σ and mean d, the α% Confidence Interval is given by

$$d \pm z_\alpha \sigma \qquad (6.19)$$

Where the values of $z_\alpha$ for different α are given in Appendix A.3.

Using the variance given in equation (6.18), the α% Confidence Interval is given by

$$d \pm z_\alpha \sqrt{\frac{E(t)[1-E(t)]}{|N_t|} + \frac{E(t-1)[1-E(t-1)]}{|N_{t-1}|}} \qquad (6.20)$$

Now, the probability that the error observed at time **t** is greater than that observed at time (t-1) is evaluated by using the equation (6.21)

$$if \left( d \geq z_{\left(1-\frac{\alpha}{2}\right)} \sqrt{\sigma^2} \right) \qquad (6.21)$$

Here, the value α = 95% and the confidence interval is considered as one-sided as it is only required to find the increase in the error rate to detect the change.

Now if the value of d is greater, then the following steps are taken:

1. The error estimated at time t is more than the average error confirming a change has occurred and the null hypothesis (H0) is rejected. Drift is said to have occurred.

2. The type of the drift occurred is detected and the size of the adaptive window is accordingly reduced till the window where the change had occurred.

3. A new classifier is trained using the labeled data. The micro clusters created are added into the ensemble.

4. The average error in the fixed section is reinitialized to the current error to reflect the current status.

If the value of d is less, then the following steps are taken:

1. The error estimated at time t is less than the average error confirming no change has occurred and the null hypothesis (H0) is accepted. Drift did not occur.

2. The adaptive window extends (grows) in size by adding the current window. If the adaptive window has reached its preset maximum size **win_size,** then the oldest window is dropped, the rest of the windows are moved forward in the

# Semi-Supervised Clustering based Classifier Model - ASSCCMI

order of their arrival time and the current window is appended. Else the current window is simply appended to the adaptive window.

3. The average error in the fixed section is recalculated to include the error statistics of the current window.
4. The ensemble model is re-trained on the misclassified instances from the current window.

The flowchart in Figure 6-9 represents the steps performed during drift detection and adaptation.

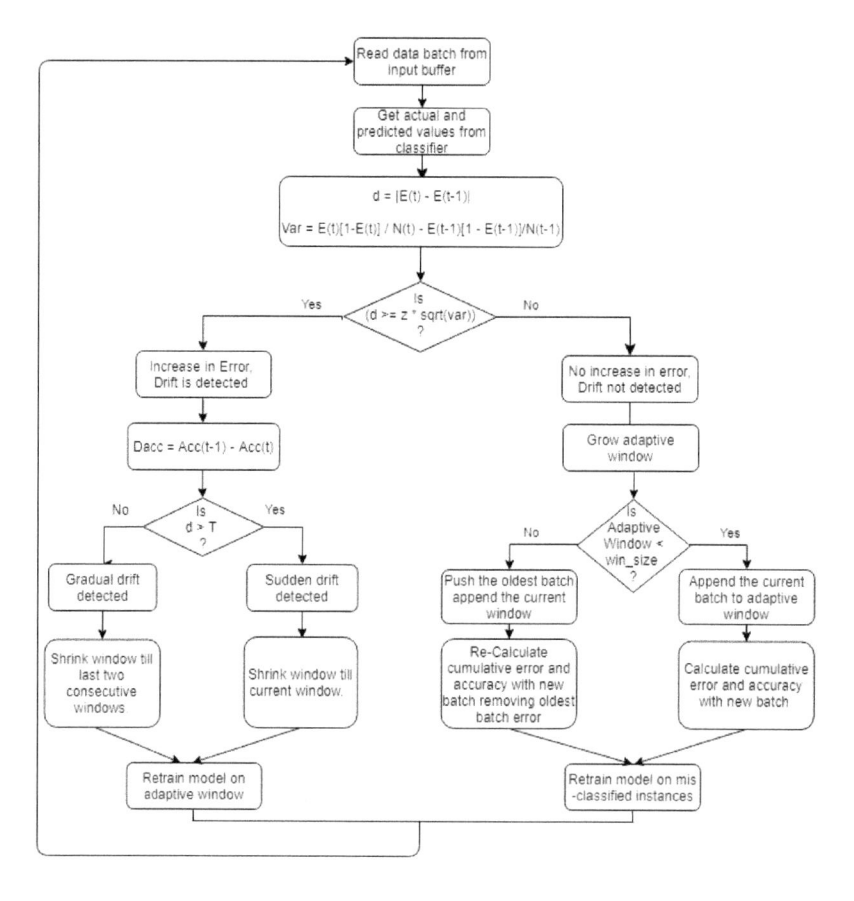

**Figure 6-9: Drift Detection and Adaptation flowchart for ASSCCMI**

## 6.7.2 Drift Adaptation

On detection of the drift, the difference between cumulative accuracy of previous batch and accuracy of current batch, $d_{acc}$, is used to determine the type of the drift. This value is compared with a threshold **T**, to check if the drift is sudden or gradual. If the value of $d_{acc}$ is higher than the threshold **T**, then there is a large difference in the error rates of the two consecutive windows thus suggesting a sudden change in the concept. The adaptive window is reduced to the current window and a new classifier model is trained on this current window. However, if the value of $d_{acc}$ is lower than the threshold **T**, then the change in the concept drift is gradual. The concept is changing incrementally. The adaptive window is shrunk to last two consecutive windows and the model is trained on data from these windows.

## 6.8 REFINING THE ENSEMBLE

The ensemble model is refined when the model is updated by adding newly trained models to it. A model is trained on the shrunk window when the drift is detected or on the misclassified instances when the drift is not detected. In both cases, Semi-supervised Clustering is invoked to create micro clusters. The micro clusters that are created are added in the ensemble. But before adding, a novel merge cluster technique to maintain the total number of micro clusters and also to balance the majority and minority micro cluster is done. The merge cluster technique is as shown in Figure 6-10.

If the total number of micro clusters exceeds the maximum number $K_M$, then the existing micro clusters in the ensemble are first merged to make space for the new ones to be added. A novel imbalance sensitive merging is applied. If a particular class label has more number of micro clusters, then the closest and the smallest micro clusters with same label are merged till their total number equals the other. This balances the number of micro clusters for all the classes. After this initial merge, if the total number of micro clusters still exceeds the threshold, then micro clusters for both labels are reduced by a number such that the total number of micro clusters is less than the threshold.

The merging is based on the distance of cluster centroids from other centroids and the total number of instances in each cluster. The shortest Euclidean distance between a pair of cluster centroids is found. Then, the new centroid is generated by taking a weighted average of the two micro cluster's centroids.

This process is repeated recursively until the total number of micro clusters is less than the threshold. The new micro clusters are then added to the ensemble.

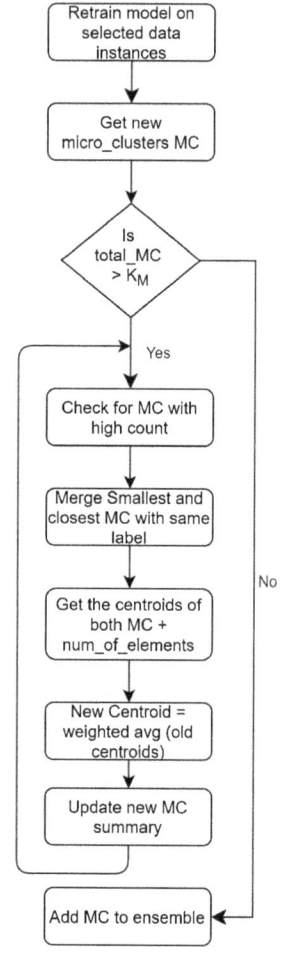

**Figure 6-10: Merge Cluster for Refining Ensemble flowchart for ASSCCMI**

The refining of ensemble aids in maintaining an updated model that is trained on the recent data stream. It also helps maintaining the size of the ensemble as well as number of clusters in each model of the ensemble.

## 6.9 SUMMARY

This chapter discussed about the proposed Semi-supervised learning model ASSCCMI, designed to solve the problem of imbalance classification in data streams when the data is scarcely labeled. The chapter discussed in detail the various modules designed for ASSCCMI. The recommended semi-supervised clustering model SSL_EM, cluster purity check module, pure micro cluster creation module, label propagation, ensemble classification, drift detection, drift adaptation, and refining are all explained in detail.

In the next chapter, the experimental setups, the datasets used for evaluating the model and evaluation metrics are explained in detail. The different stream classification algorithms and the reason for choosing KNNADWIN and OzaBaggingADWIN for comparing with ASSCCMI are also explained. The parameter settings used for evaluating ASSCCMI as well as stream classification algorithms are also discussed in the next chapter.

# Chapter 7

Experimental Setup for ASSCCMI

# 7. EXPERIMENTAL SETUP FOR ASSCCMI

In order to evaluate the performance of the proposed research model ASSCCMI, experiments were performed on a number of datasets. In this section, the datasets used to perform the experiments and evaluate the proposed model are introduced. This is followed by a discussion on the different metrics used for evaluating the performance of ASSCCMI. To evaluate the performance, ASSCCMI is compared with two of the standard state-of-art stream classification algorithms viz., KNNADWIN and OzaBaggingADWIN. ASSCCMI is also compared with a recent semi-supervised model SPASC [93] which has outperformed all its predecessors. These algorithms are also explained in this section. The various parameters and their tuning are then explained in the following sub section.

## 7.1 EXPERIMENTAL SETUP

The experiments are performed on a Windows 10 computer using an $5^{th}$ Generation Intel Core i5-5200U CPU with 2.20 GHz processor speed and 8GB RAM.

The model HECMI, CDACI as well as ASSCCMI are all implemented using Python 3.7.3 with pandas for working with data frames, numpy for arrays and sklearn metrics packages for evaluating the algorithms. Standard algorithms of data streams viz., KNNADWIN and OzaBaggingADWIN were implemented using the scikit-multiflow packages [111] available in Python 3.5 and above.

## 7.2 DATASET DESCRIPTION

The proposed model is evaluated on some of the benchmark datasets which have been used in various researches especially in data streams. Three (03) real datasets viz., Electricity [99] [5], Spambase [100], Credit Card Fraud Detection [100] and four (04) synthetically generated data viz., Hyperplane 1 & Hyperplane 2 [98] and SEA1 & SEA2 [52] are used for evaluation. The synthetic datasets are generated using the MOA - Massive Online Analysis [112]. MOA is a software environment for implementing algorithms and running experiments for online learning from evolving data streams. MOA includes a collection of offline and online methods as well as tools for evaluation.

### 7.2.1 Electricity dataset

The Electricity is one of the most widely used dataset for concept drift experiments. This dataset was described by M. Harries and analyzed by J. Gama. The data represents the electricity market prices which are affected by the demand and supply of the market. It includes the data about electricity transfer between New South Wales and Victoria states of Australia. The recordings of data are set for every five minutes. The Electricity transfers to/from the neighbouring state of Victoria were done to meet the demand and remove fluctuations.

The total number of instances in the dataset is 45312 instances and was collected over a period dated from 7$^{th}$ May 1996 to 5$^{th}$ December 1998. Each instance of the dataset refers to a period of 30 minutes, so for a whole days' recording there are 48 instances. There are total 8 attributes where the 7 attributes are the day of week, the time stamp, the New South Wales electricity demand, the Victoria electricity demand, the scheduled electricity transfer between states and 8$^{th}$ attribute is the class label. The class label identifies the change of the price (UP or DOWN) in New South Wales relative to a moving average of the last 24 hours. The levels UP/DOWN represent deviations of price on a one day average basis. The attribute information and class distribution for Electricity dataset is described in Table 7-1 below:

**Table 7-1: Attribute information for Electricity dataset**

| Total Attributes | Class Label | Class Distribution |
|---|---|---|
| 8 | 0 – DOWN, 1 – UP | 0 – 26075 (57.55%) 1 – 19237 (42.45%) |

### 7.2.2 Spambase dataset

The "spam" includes diverse concepts such as advertisements for products, fast money schemes, chain letters etc. The spam e-mails contains a collection of emails from the postmaster and individuals who had filed for spam whereas the non-spam e-mails came from filed work and personal e-mails.

The dataset contains 4601 instances with 58 attributes. All 57 attributes are continuous attribute. The 58$^{th}$ attribute is the class label that depicts an email is 'spam' or 'not spam'. Most of the attributes indicate whether a particular word or character was frequently occurring in the email. The run-length attributes (55-57) measure the length of sequences of consecutive capital letters. The attribute information and class distribution for Spambase dataset is described in Table 7-2 below:

Table 7-2: Attribute information for Spambase dataset

| Total Attributes | Class Label | Class Distribution |
|---|---|---|
| 58 | 0 – ham, 1 – spam | 0 – 2788 (60.6%) 1 – 1813 (39.4%) |

### 7.2.3 Credit Card dataset

The dataset contains information about German credit data that classifies people as having a good or bad credit risks. The dataset contains 1000 instances with 19 attributes and a class label with value 1 or 0. The dataset contains categorical as well as numeric attributes. This research uses the dataset with all numeric attributes. The attribute information and class distribution for Credit Card dataset is described in Table 7-3 below:

Table 7-3: Attribute information for CreditCard dataset

| Total Attributes | Class Label | Class Distribution |
|---|---|---|
| 21 | 0 – bad, 1 – good | 0 – 300 (30%) 1 – 700 (70%) |

### 7.2.4 SEA dataset

The SEA – **S**treaming **E**nsemble **A**lgorithm dataset consists of three attributes and a class attribute. The values of the three attributes lie between 0 and 10. The class consists of two labels and the decision boundary is given by $f1 + f2 \leq \alpha$, where $f1$ and $f2$ are the first two attributes and $\alpha$ is a threshold value. The most frequent values of $\alpha$ are 7, 8, 9 and 9.5. The SEA generator from the MOA framework is used, to generate two datasets viz., SEA1 and SEA2, each with 100,000 instances and 3 attributes (refer Appendix A.1). SEA1 is induced with a gradual drift and SEA2 with sudden drift. In SEA1, Gradual drift is induced by changing the function from $f2$ to $f3$ over a window of 20000 instances. In SEA2, Sudden drift is also induced by changing the function from $f2$ to $f3$ at $50000^{th}$ instance. The attribute information for SEA1 and SEA2 are described in Table 7-4 and Table 7-5 below:

**Table 7-4: Attribute information for SEA1 dataset**

| Total Attributes | Class Label | Class Distribution |
|---|---|---|
| 4 | 0 – GroupB, 1 – GroupA | 0 – 63945 (63.94%) 1 – 36055 (36.06%) |

**Table 7-5: Attribute information for SEA2 dataset**

| Total Attributes | Class Label | Class Distribution |
|---|---|---|
| 4 | 0 – GroupB, 1 – GroupA | 0 – 63963 (63.96%) 1 – 36037 (36.04%) |

### 7.2.5 Moving Hyperplanes dataset

The moving hyperplanes dataset is used for modelling the problem of predicting the label of rotating hyperplane. A hyperplane is the set of points $x_i$ with direction $w_i$ in a d dimensional plane. A decision boundary of the hyperplane is denoted by $\sum x_i . w_i = 0$. Instances with $\sum x_i . w_i > 0$ are labeled as positive while instances with $\sum x_i . w_i < 0$ are labeled as negative. The hyperplane generator from the MOA framework [112] is used to generate Hyperplane1 and Hyperplane2 with 100,000 instances and 10 attributes. Drifts are simulated by changing the orientation and position of the planes by varying the values of weights as the stream advances. Hyperplane1 contains incremental drifts and Hyperplane2 contains subtle incremental drifts (refer Appendix A.1). Incremental drifts are induced by changing the magnitude of weights by a factor of 0.01 and with a probability of 10 to reverse the orientation. Similarly, for subtle incremental drifts the magnitude of weights is changed by a factor of 0.001 with same probability to reverse the orientation. The attribute information for Hyperplane1 and Hyperplane2 are described in Table 7-6 and Table 7-7 below:

Experimental Setup for ASSCCMI

Table 7-6: Attribute information for Hyperplane1 dataset

| Total Attributes | Class Label | Class Distribution |
|---|---|---|
| 11 | 0 – Class1, 1 – Class2 | 0 – 50516 (50.52%) 1 – 49484 (49.48%) |

Table 7-7: Attribute information for Hyperplane2 dataset

| Total Attributes | Class Label | Class Distribution |
|---|---|---|
| 11 | 0 – Class1, 1 – Class2 | 0 – 50192 (50.19%) 1 – 49808 (49.81%) |

### 7.2.6 Summary of Datasets

Detailed dataset descriptions are mentioned in Appendix A.2. The Table 7-8 below summarizes the characteristics to the datasets described above.

Table 7-8: Summary of datasets used for evaluating ASSCCMI

| Dataset Name | Real / Synthetically generated | Total No. of instances | Imbalance ratio (IR) | Nature of data |
|---|---|---|---|---|
| Hyperplane1 | Synthetic | 100000 | 1.01 | Incremental drift |
| Hyperplane2 | Synthetic | 100000 | 1.00 | Subtle incremental drift |
| SEA1 | Synthetic | 100000 | 1.79 | Imbalance with gradual drift |
| SEA2 | Synthetic | 100000 | 1.32 | Imbalance with sudden drift |
| Electricity | Real | 45312 | 1.36 | Imbalance with incremental |
| Spam base | Real | 4601 | 1.54 | Imbalance |
| Credit Card | Real | 1000 | 2.33 | Imbalance |

The imbalance ratio (IR) for each dataset is calculated as follows:

$$IR = \frac{Number\ of\ majority\ class\ instances}{Number\ of\ minority\ class\ instances} \quad (7.1)$$

## 7.3 EVALUATION METRICS

Evaluation metrics are the most important aspect for a classifier that acts as an indicator of the performance of the model. Most data mining algorithms do not account for the underlying class-imbalance. The standard measures viz., error rate and accuracy are inefficient in dealing with the imbalance problem due to their bias towards majority class.

For a two class problem, confusion matrix is calculated as shown below:

**Table 7-9: Confusion matrix**

| Actual Class | | Predicted Class | |
|---|---|---|---|
| | | Positive | Negative |
| | Positive | True Positive (TP) | False Negative (FN) |
| | Negative | False Positive (FP) | True Negative (TN) |

The terms in the confusion matrix are defined as follows [7]:

1. True positives: The positive tuples that were correctly classified by the classifier i.e. positive class labeled as positive.

2. True negatives: The negative tuples that were correctly classified by the classifier i.e. negative class labeled as negative.

3. False positives: The negative tuples that were incorrectly labeled as positive i.e. the negative class labeled as positive.

4. False negatives: The positive tuples that were incorrectly labeled as negative i.e. the positive class labeled as negative.

The confusion matrix also helps to measure other important performance measures related to the classifier such as accuracy, sensitivity, specificity, precision and recall.

**Accuracy**: The percentage of tuples correctly classified by the classifier i.e. it represents how often the classifier predicts correct outcome. It is defined as:

$$Accuracy = \frac{TP + TN}{TP + TN + FP + FN} \quad (7.2)$$

**Precision**: It determines the total percentage of instances that the classifier labeled as minority actually belongs to the minority class. Precision is also referred to as measure of exactness. It basically specifies out of the classified minority class, how many actually belong to minority class.

$$Precision = \frac{TP}{TP + FP} \quad (7.3)$$

**Recall**: It determines the percentage of minority class instances that are labeled correctly as minority. Recall is also called as a measure of completeness. It basically specifies out of the total minority class how many were correctly classified. For imbalance classes, the recall value should be higher. It then suggests that the classifier can correctly classify minority class instances.

$$Recall = \frac{TP}{TP + FN} \quad (7.4)$$

For a learning model, the objective should be increasing the recall but without affecting precision. A classifier can achieve high precision by correctly classifying all positive tuples as positive but at the same time it achieves a lower recall if it misclassifies many other tuples as positive. F-measure is defined to show the trade-off between them.

**AUC:** The area under ROC Curve (Receiver Operating Characteristics) is plotted as True Positive rate vs the False Positive rate. It aides in visualizing the trade-off between the rate at which the model can accurately recognize positive cases versus the rate at which it mistakenly identifies negative cases as positive for different portions of the test set. Any increase in TPR occurs at the cost of an increase in FPR. The area under the ROC curve is a measure of the accuracy of the model.

## 7.4   STANDARD ALGORITHMS FOR PERFORMANCE EVALUATION

The performance of ASSCCMI is compared with two of the standard stream classification algorithms which are capable of detecting concept drift and adapt to the change and it is also compared with an existing semi-supervised classification framework which is used to classify scarcely labeled data streams with concept drift and is found to resemble closest ASSCCMI in its functioning.

### 7.4.1 Stream Algorithms for Comparison

Many traditional learning algorithms have been extended or modified to make them suitable for the data streams environment. Apart from simple drift detection techniques like DDM, EDDM, ADWIN etc., there are online versions of state-of-art classifiers like the Naïve Bayes classifiers, Decision trees and also Perceptron networks, which perform in the online environment [13]. The stream classification also consists of lazy learners and ensemble methods that perform classification on data streams [112].

The lazy learner methods listed in Table 7-10 include the KNNClassifier, KNNADWINClassifier, SAMKNNClassifier and KNNRegressor.

**Table 7-10: Lazy learning methods of Stream Classification**

| | |
|---|---|
| KNNClassifier | k-Nearest Neighbors classifier. |
| KNNADWINClassifier | k-Nearest Neighbors classifier with ADWIN change detector. |
| SAMKNNClassifier | Self Adjusting Memory coupled with the kNN classifier. |
| KNNRegressor | k-Nearest Neighbors regressor. |

The KNNClassifier is a basic non-parametric model that classifies a data stream by finding k closest neighbours to the query sample in the data sample. Euclidean distance is used to find the k-neighbours. KNNClassifier does not perform any explicit drift detection. KNNADWINClassifier uses KNNClassifier as a base classifier and ADWIN for change detection. The SAMKNNClassifier i.e. Self-Adjusting Memory - also uses KNNClassifier as its base classifier. It creates an ensemble of the classifiers that can target a current or a former concept. It builds two memories: STM for current concept and LTM for retaining information of a former concept. The KNNRegressor is a non-parametric regressor that keeps track of the latest window training samples. It performs prediction using the k-nearest neighbour stored samples.

The ensemble methods include varied ensemble of classifiers that are based on under and over sampling strategies applied for the online classification. Some of them that focus on drift detection are listed in the Table 7-11 below:

# Experimental Setup for ASSCCMI

**Table 7-11: Ensemble learning methods of Stream classification**

| | |
|---|---|
| OnlineAdaC2Classifier | Online AdaC2 ensemble classifier. |
| OnlineBoostingClassifier | Online Boosting ensemble classifier. |
| OnlineCSB2Classifier | Online CSB2 ensemble classifier. |
| OnlineRUSBoostClassifier | Online RUSBoost ensemble classifier. |
| OnlineSMOTEBaggingClassifier | Online SMOTEBagging ensemble classifier. |
| OnlineUnderOverBaggingClassifier | Online Under-Over-Bagging ensemble classifier. |
| OzaBaggingClassifier | Oza Bagging ensemble classifier. |
| OzaBaggingADWINClassifier | Oza Bagging ensemble classifier with ADWIN change detector. |

The online classifiers are online extensions of their traditional classification methods. OnlineAdaC2Classifier is the adaptation of ensemble learner to data streams. OnlineBoostingClassifier is the extension of AdaBoost classifier for online classification. OnlineCSB2Classifier is the online version of the ensemble learner CSB2. CSB2 algorithm is a hybrid AdaBoost and AdaC2. It treats the misclassified instances like AdaC2 and correctly classified examples AdaBoost. Online RUSBoost follows the techniques of UnderOverBagging and SMOTEBagging by introducing sampling techniques as a post-processing step that is performed before iteration of the standard AdaBoost algorithm. Online SMOTEBagging is the online version of the ensemble method SMOTEBagging where the negative class instances are re-sampled with 100% replacement. OnlineUnderOverBagging is the online version of the ensemble method. For imbalanced classes, it does under-sampling of majority class and over-sampling of minority class. The diversity of the base learners are also changed by varying the sampling rate. The OzaBaggingClassifier is an improvement over the well-known Bagging ensemble. In this method, each training sample drawn by a binomial distribution is trained k times. The OzaBaggingADWINClassifier is an improvement over the OzaBaggingClassifier which comes due to the addition of ADWIN change detector. The ADaptive WINdow keeps statistics of a variable size window to detect change whenever the statistics changes above a threshold.

**KNNADWIN** from the lazy learner method and **OzaBaggingADWIN** from the ensemble method are used for evaluating the performance of ASSCCMI. Both the methods use adaptive windows to detect and adapt to changes similar to ASSCCMI.

KNNADWIN uses k nearest neighbours to find closest neighbours and ASSCCMI also performs clusters using similarity between closest elements. OzaBaggingClasssifier is an ensemble classifier and ASSCCMI is also an ensemble of cluster based classifiers. These two methods have techniques similar to ASSCCMI. Hence they are selected for evaluating the performance of ASSCCMI. The algorithms for KNNADWIN and OzaBaggingADWIN are presented in the Appendix A.4 and A.5 respectively.

KNNADWIN and OzaBaggingADWIN are basically supervised learning methods and ASSCCMI is a semi-supervised learning model. Hence for a complete evaluation of the performance of ASSCCMI, a closely resembling semi-supervised learning based classification model SPASC is chosen from the literature referred.

### 7.4.2 Semi-Supervised Learning Algorithm for Comparison

Among the different framework discussed in Table 2-5, it is observed that SPASC is one of the best semi-supervised classification methods proposed in the field of non-stationary data streams. It overcomes the drawbacks of its predecessors and has shown to improve performance for the datasets it is tested on. Like ASSCCMI, SPASC makes use of ensemble of cluster based classifiers. ASSCCMI, however, is focused on imbalanced data stream classification. Unlike SPASC, ASSCCMI does not assume any distribution of data (e.g Gaussian, Poisson etc) while training the ensemble, thus making it more suitable for real time scenarios. ASSCCMI also overcomes the problem of poor minority recall of SPASC, by first creating pure clusters and then performing similarity based label propagation for unlabeled clusters. Unlike SPASC, ASSCCMI uses informed drift detection and adaptation technique using adaptive windows and statistical hypothesis testing.

### 7.5  PARAMETER TUNING

The different parameters required in the efficient working of the model are decided after thorough experimentation. Some parameters like the number of models in the ensemble, ADWIN window size are kept constant for all the datasets. There are some of the parameters which are dataset dependent and are set accordingly. The parameter settings for ASSCCMI are explained first followed by the parameter setting for KNNADWIN, OzaBaggingADWIN and SPASC.

## 7.5.1 Parameter Tuning for ASSCCMI

The default parameter settings are as follows:

(i) The initial number of initial cluster K = 10 or 12 for each model in the ensemble. Total number of micro-cluster number $K_M \leq 100$ for all datasets;

(ii) The number of models in an ensemble i.e. M = 6

(iii) The batch_size is different for every datasets and is finalized after experimentation.

(iv) The input buffer is kept to be twice the batch size and

(v) The maximum size for ADWIN to grow is win_size = 6*batch_size.

(vi) Convergence threshold $\varepsilon = 10^{-4}$

(vii) The threshold for differentiating between sudden and gradual drift T = 0.045

Table 7-12: Parameter tuning for ASSCCMI

| Dataset Name | Total No. of instances | Batch_size | Initial no. of clusters k | No. of models in ensemble |
|---|---|---|---|---|
| Hyperplane1 | 100000 | 1000 | 12 | 6 |
| Hyperplane2 | 100000 | 1000 | 12 | 6 |
| SEA1 | 100000 | 1000 | 12 | 6 |
| SEA2 | 100000 | 1000 | 12 | 6 |
| Electricity | 45312 | 1000 | 12 | 6 |
| Spambase | 4601 | 100 | 10 | 6 |
| CreditCard | 1000 | 100 | 10 | 6 |

### 7.5.2 Parameter Tuning for Comparative algorithms

Parameters of KNNADWIN classifier and their settings are listed in the Table 7-13 below. OzaBaggingADWIN uses KNNADWIN classifier as its base estimator.

**Table 7-13: Parameter tuning for stream algorithms**

| Parameters | Description | Parameter value |
|---|---|---|
| n_neighbors | The number of nearest neighbors to search for | 10 |
| max_window_size | The maximum size of the window storing the last viewed samples | 3000 |
| leaf_size | The maximum number of samples that can be stored in one leaf node | 1000 |
| metric | The distance metric to use for the KDTree | Euclidean |
| n_estimator (only for OzaBaggingADWIN) | The size of the ensemble | 6 |

Parameters of SPASC are kept same as suggested by the authors in [93]. These include the number of clusters in a model $Q = 5$. Batch size for each dataset is kept similar to the size used for ASSCCMI.

### 7.5.3 Screenshots of Experimentation

In this section, the working of the comparative algorithms KNNADWIN, OzaBaggingADWIN and SPASC as well as the ASSCCMI model for the Credit Card dataset is shown. The Figure7-1 (a) – (c) depicts the execution and the final result of KNNADWIN classifier on Credit Card dataset.

Figure 7-1 (a) is the python code for KNNADWIN, Figure 7-1 (b) is the execution results window and Figure 7-1 (c) is the csv file storing the results of KNNADWIN for the Credit Card dataset.

# Experimental Setup for ASSCCMI

(a)

(b)

# Experimental Setup for ASSCCMI

(c)

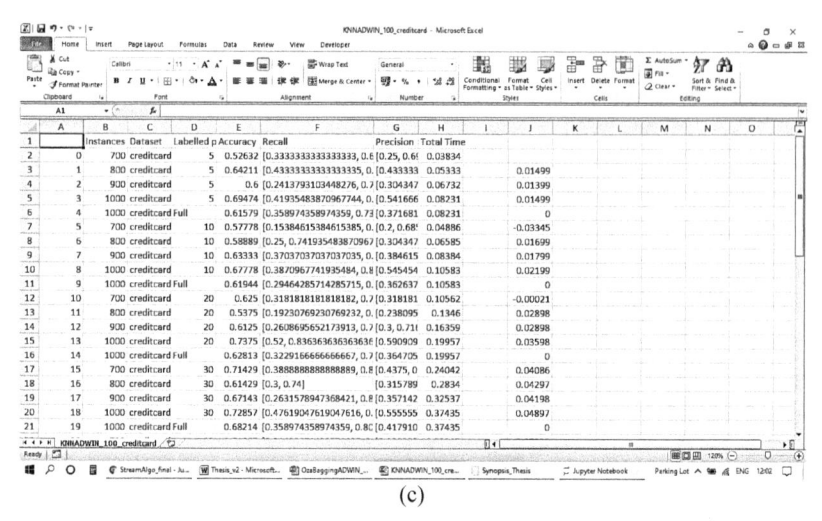

**Figure 7-1: KNNADWIN for Credit Card dataset**

The Figure 7-2 (a) – (c) depicts the execution and the final result of OzaBaggingADWIN classifier on Credit Card dataset. Figure 7-2 (a) is the python code for OzaBaggingADWIN, Figure 7-2 (b) is the execution results window and Figure 7-2 (c) is the csv file storing the results of OzaBaggingADWIN for the Credit Card dataset.

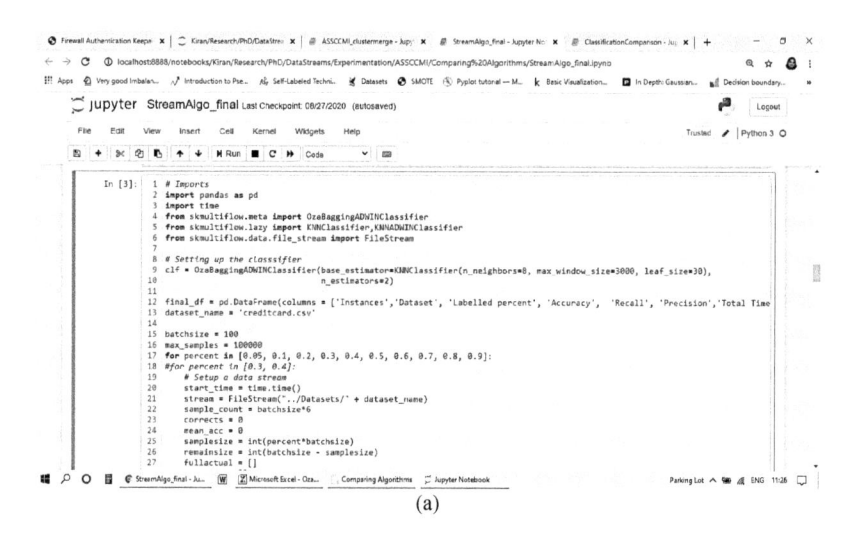

(a)

# Experimental Setup for ASSCCMI

(b)

(c)

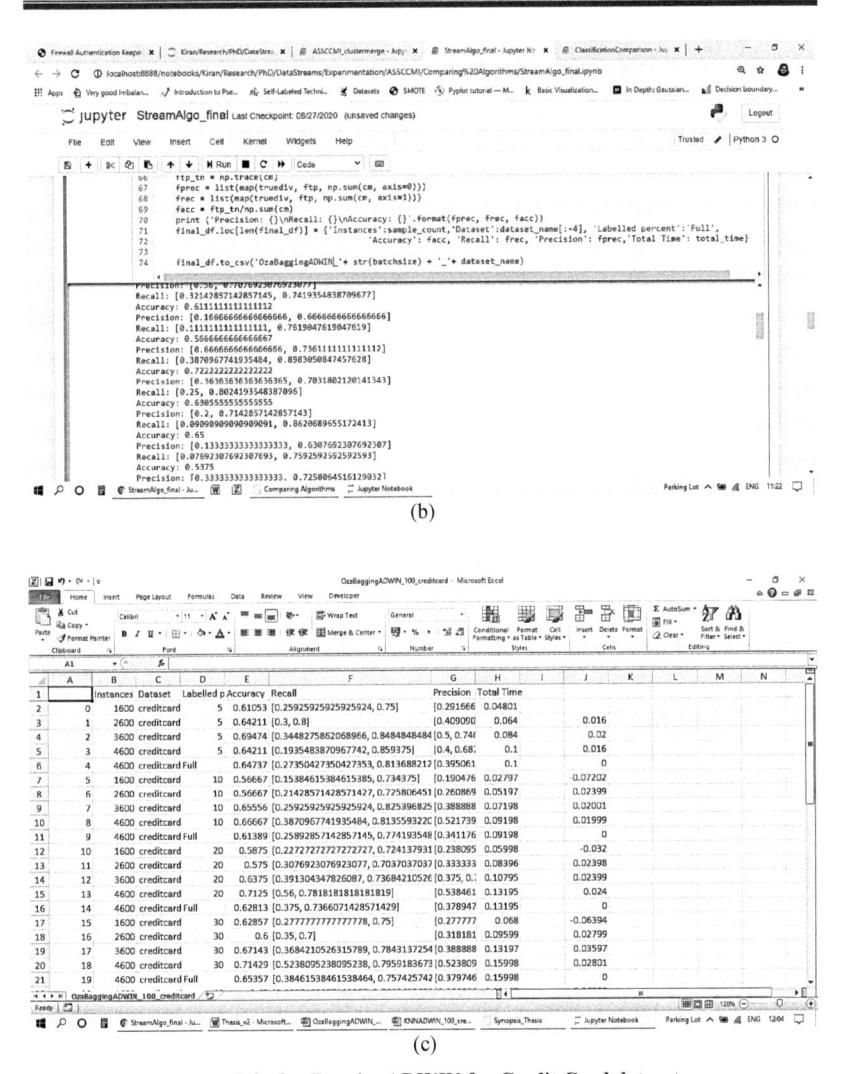

**Figure 7-2: OzaBaggingADWIN for Credit Card dataset**

The Figure 7-3 (a) – (c) depicts the execution and the final result of SPASC classifier on Credit Card dataset. Figure 7-3 (a) is the python code for SPASC, Figure 7-3 (b) is the execution results window and Figure 7-3 (c) is the csv file storing the results of SPASC for the Credit Card dataset.

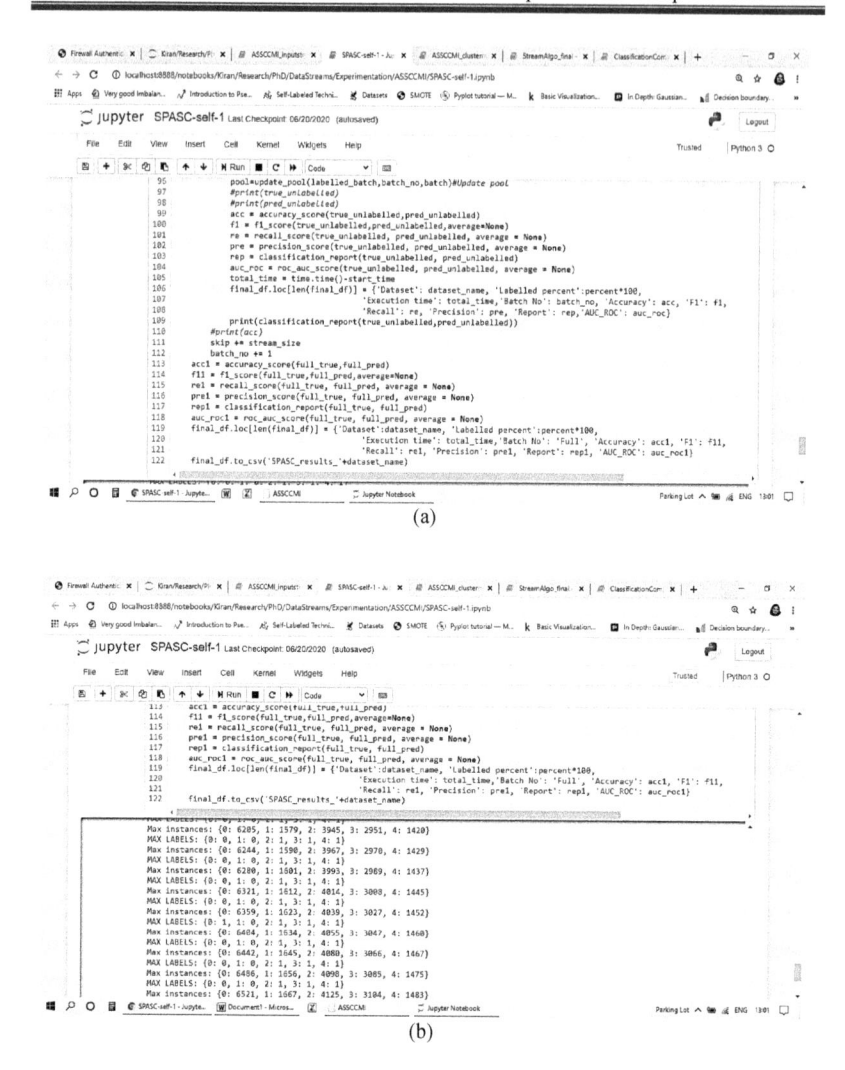

# Experimental Setup for ASSCCMI

(a)

(b)

Experimental Setup for ASSCCMI

(c)

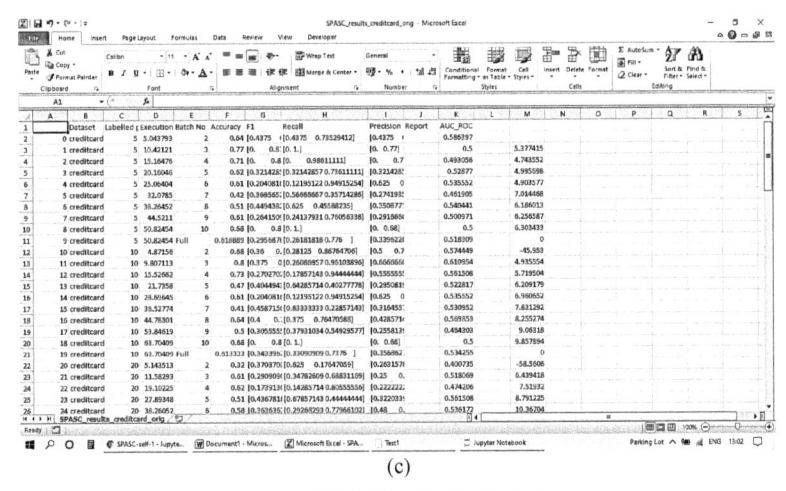

Figure 7-3: SPASC for Credit Card dataset

The Figure7-4 (a) – (c) depicts the execution and the final result of ASSCCMI classifier on Credit Card dataset. Figure 7-4 (a) is the shows the state of the input stream module where the batch is not ready and the module is reading from the stream source. Figure 7-4 (b) both the batch and model are ready and execution results are shown and Figure 7-4 (c) is the result csv file storing the results of ASSCCMI for the Credit Card dataset.

(a)

120

Experimental Setup for ASSCCMI

(b)

(c)

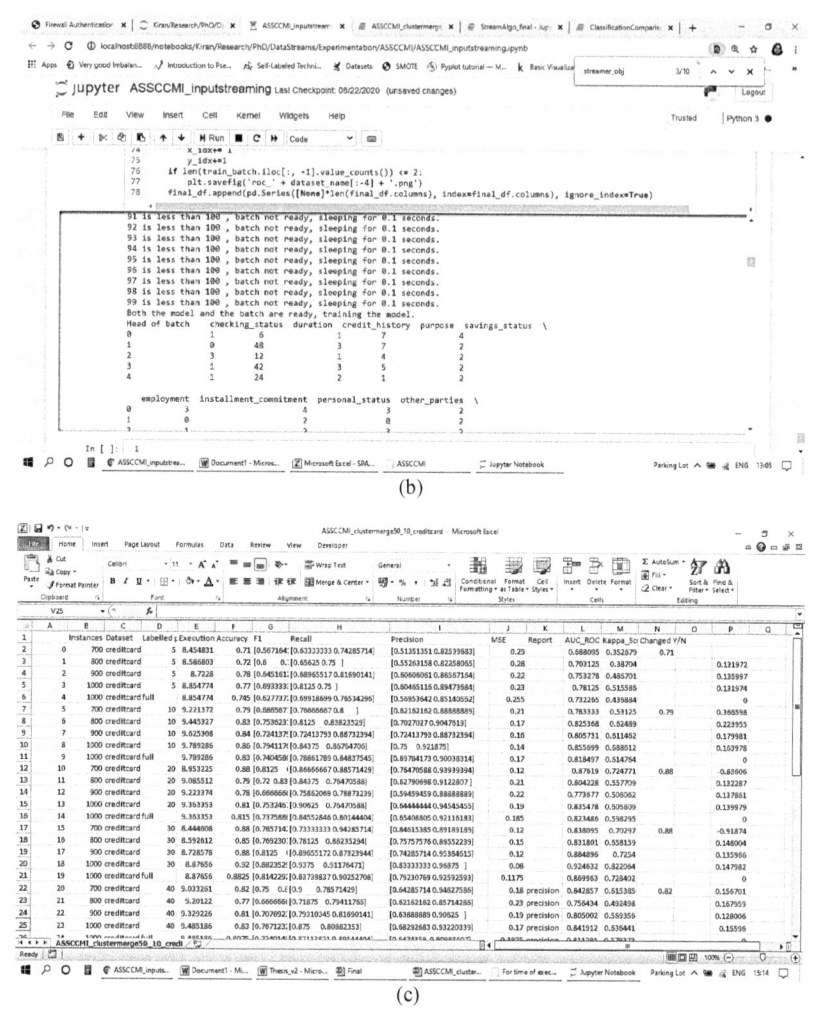

Figure 7-4: ASSCCMI for Credit Card dataset

## 7.6 SUMMARY

This chapter explained the experimental setup, datasets, comparative algorithms, parameter tuning, experimentation screenshots and evaluation metric. In the next chapter, test plan, test cases and experiments performed to evaluate the proposed framework are discussed. A detail result analysis on all the test cases for all the data sets under consideration is also explained.

# Chapter 8

Research Findings and Analysis

# 8. RESEARCH FINDINGS AND ANALYSIS

The research objectives described in Section 3.1 were all achieved through the adaptive framework ASSCCMI described in Chapter 6. To verify that the research objectives were successfully attained, test cases were designed and the performance of the model was evaluated. Tests are performed to evaluate the functionality of modules in the ASSCCMI. The Test plan and test cases are designed to evaluate the ASSCCMI as a whole as well as its modules and sub-modules. In section 8.1 the description of the test plan and its respective test cases is done followed by the mapping of test cases with the research objectives. In section 8.2 the test cases and the test hypothesis are designed and evaluated.

## 8.1 TEST PLAN AND TEST CASES

ASSCCMI is evaluated to verify the performance of the model and framework on various aspects like accuracy with imbalanced data, drift detection, accuracy in presence of drifts (drift adaptation), overall accuracy and execution time. The Test Plan included designing Test Cases for evaluating the model in parts. The parts and their respective test cases are as discussed below:

**PART I:** This part includes testing the Semi-Supervised Clustering based Classification Module (SSC) of ASSCCMI. The test cases aid in evaluating the classification module SSC. The proposed SSL_EM is evaluated first to check if it is able to learn and classify using scarcely labeled data. This evaluation is done using the Test Case 1. The next part is to check if the cluster purity check module improves the performance of the SSL_EM module. The test case 2 evaluates the performance of SSC when the purity check module is added.

1. **Test Case 1: To assess the proposed Semi-Supervised Clustering SSL_EM approach**

    a. TC1.1: To evaluate the accuracy of proposed SSL_EM

    b. TC1.2: To evaluate the AUROC values for proposed SSL_EM

2. **Test Case 2: To assess the cluster purity check module for performance improvement of SSC**

# Research Findings and Analysis

**PART II:** This part involves testing the Concept Drift Detection and Adaptation Module of ASSCCMI. These test cases help in verifying the drift detection and adaptation capability of the proposed model. The test case 3.1 evaluates if the ASSCCMI is able to detect the drifts correctly. The test case 3.2 evaluates if the ASSCCMI is able to adapt to the drifts by evaluating the improvement in accuracy with drift detection.

3. **Test Case 3: To analyze the drift detection and adaptation efficacy of ASSCCMI**
   a. TC3.1: To evaluate drift detection and adaptation capability of ASSCCMI
   b. TC3.2: To evaluate the overall accuracy of ASSCCMI with and without drift detection and adaptation on data streams

**PART III:** This part deals with testing the overall ASSCCMI. These test cases test the overall performance of the proposed model. The improvement in learning is evaluated by comparing the performance of the model in terms of accuracy with different stream classification techniques and semi-supervised learning technique. The test case 4 evaluates the improvement in accuracy of ASSCCMI for different data streams. The improvement in learning in terms of minority class recall is evaluated in test 5 and the improvement in execution time is evaluated in the test case 6.

4. **Test Case 4: To evaluate the overall accuracy of ASSCCMI**
   a. TC4.1: To evaluate the overall accuracy of ASSCCMI on imbalanced data streams
   b. TC4.2: To evaluate the overall accuracy of ASSCCMI on balanced data streams with concept drifts
   c. TC4.3: To evaluate the overall accuracy of ASSCCMI on imbalanced data streams with concept drifts

5. **Test Case 5: To evaluate the overall performance in terms of minority class recall of ASSCCMI in presence of imbalanced data**

6. **Test Case 6: To assess the execution time of ASSCCMI**

# Research Findings and Analysis

The above test cases were designed to verify the attainment of the research objective and justifying the proposed design ASSCCMI. The mapping of test cases (TC) to the research objective (RO) is as shown in Table 8-1 below:

Table 8-1: Mapping of Research Objectives and Test Cases

|        | Test Case 1 | Test Case 2 | Test Case 3 | Test Case 4 | Test Case 5 | Test Case 6 |
|--------|:---:|:---:|:---:|:---:|:---:|:---:|
| RO 1.a |   | ✓ | ✓ | ✓ | ✓ | ✓ |
| RO 1.b |   |   |   | ✓ | ✓ | ✓ |
| RO 1.c | ✓ |   |   | ✓ | ✓ |   |
| RO 2.a | ✓ |   |   | ✓ | ✓ |   |
| RO 2.b | ✓ | ✓ |   | ✓ | ✓ | ✓ |
| RO 2.c | ✓ | ✓ |   | ✓ | ✓ |   |
| RO 3.a |   |   | ✓ | ✓ |   | ✓ |
| RO 3.b |   |   | ✓ | ✓ |   | ✓ |
| RO 3.c |   |   | ✓ | ✓ |   | ✓ |
| RO 3.d |   |   | ✓ | ✓ | ✓ | ✓ |
| RO 4.a | ✓ | ✓ | ✓ | ✓ | ✓ | ✓ |

The attainment of the research objectives are evaluated through execution of different hypothesis and the metrics used for evaluation are primarily accuracy and recall. Accuracy provides the best measure of performance for any classification task. In the presence of imbalanced data, accuracy does not provide the accurate measure of the correct classification of minority class data. Hence the use of recall as a metric is recommended. In addition to this, throughout the literature studied on this topic, these two metrics are majorly used.

## 8.2 TEST CASES AND TEST HYPOTHESIS

The experimental setup and parameter tuning for all the test cases and experiments is as described in section 7.1 and section 7.5 respectively. The different evaluation metrics used are also explained in section 7.3. The ASSCCMI and SSL_EM are compared with state-of-art techniques that are discussed in section 7.4. The evaluation of the proposed ASSCCMI is performed in the following way:

1. For each test case firstly an experimental setup is explained.
2. Then the results of the experiments are depicted with the help of appropriate graphs and result table.

3. Lastly an analysis on each of the result graphs / table is performed.

### 8.2.1 Testing the Semi-Supervised Clustering approach SSL_EM

The semi-supervised clustering approach SSL_EM is evaluated for its performance w.r.t to accuracy and AUROC values. The SSL_EM is evaluated to check if it is able to learn from the labeled and unlabeled data. The SSL_EM is also evaluated to check if it is able to distinguish between minority and majority classes to a satisfactory level using AUROC values. Two hypotheses are defined for this purpose. Hypothesis 1 is designed to evaluate the performance in terms of accuracy and Hypothesis 2 is designed to evaluate the AUROC values.

*8.2.1.1 Test for Accuracy of proposed SSL_EM*

***Hypothsis 1:*** SSL_EM performs better in terms of accuracy than the standard clustering and ensemble classification technique.

**Experimental Setup:**

To test this hypothesis, SSL_EM was executed on different datasets that are described in section 7.2. The accuracy of the approach was compared with the accuracy of state-of-art clustering techniques like k-means random, k-means++ and GMM. The semi-supervised clustering approach uses extension of EM for semi-supervised learning and EM basically is an improvement of k-means algorithm. Hence, from the existing standard clustering techniques, k-means random, k-means++ and GMM algorithms are selected for comparison of performance. SSL_EM is an ensemble based classifier hence, along with the standard clustering techniques; its performance is also compared with an ensemble classifier. Random initializations were used for the clustering techniques. While classifying, the clustering techniques were trained on first 6 data batches and the result was evaluated for the next batches. The ensemble classifier designed with Naïve Bayes as the base classifier is also used for this evaluation. Similar to SSL_EM, the ensemble is also trained on 6 batches for 6 initial models. Max voting strategy is then used to predict the unseen batch. For each dataset, it was assumed that only p% labeled data was available. And each technique was evaluated on the same p% labeled batch. The SSL_EM was evaluated for 5, 10, 20, 50 and 80 percent labeled data. For testing

SSL_EM, the stream input module was not considered; datasets were directly split into batches of equal size.

**Analysis:**

The following graphs show the accuracy values computed on different datasets using the aforementioned clustering and ensemble classification techniques. For each of the graphs in this test case, the X-axis depicts the different percent of labeled data that is available in each batch and the Y-axis depicts the overall accuracy for the respective percent of labeled data. The accuracy values lie in the range of [0 – 1].

It is observed that the proposed semi-supervised clustering approach works well for almost all the datasets on different percent labeled data. The accuracy values of SSL_EM are better than and quite higher as compared to the different techniques especially the clustering techniques. It was also observed that the performance of SSL_EM was consistent over the number of times the model is executed. However, for the clustering techniques the results varied largely due to random initialization of the methods.

**Spambase dataset:**

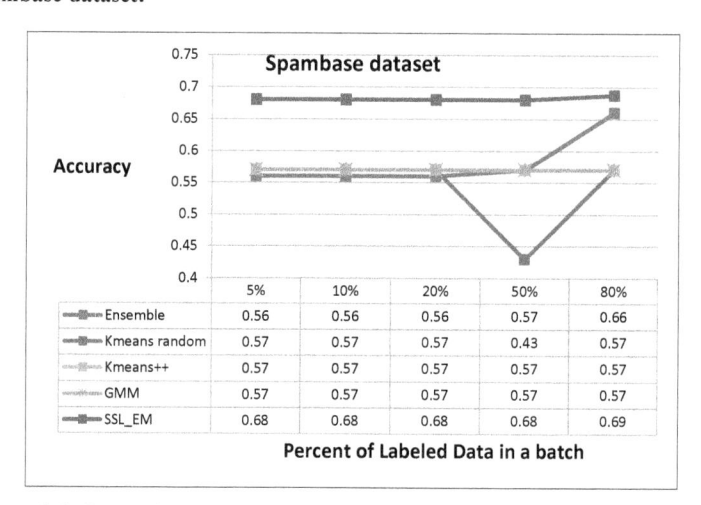

| | 5% | 10% | 20% | 50% | 80% |
|---|---|---|---|---|---|
| Ensemble | 0.56 | 0.56 | 0.56 | 0.57 | 0.66 |
| Kmeans random | 0.57 | 0.57 | 0.57 | 0.43 | 0.57 |
| Kmeans++ | 0.57 | 0.57 | 0.57 | 0.57 | 0.57 |
| GMM | 0.57 | 0.57 | 0.57 | 0.57 | 0.57 |
| SSL_EM | 0.68 | 0.68 | 0.68 | 0.68 | 0.69 |

Figure 8-1: Comparison of Accuracy of SSL_EM v/s other algorithms for Spambase

Research Findings and Analysis

1. For Spambase dataset as shown in Figure 8-1, the performance of SSL_EM is consistent for all the labeled data. GMM, k-means++ and k-means random also has consistent values of accuracy for all the labeled data but these have lower accuracy than SSL_EM. K-means random is quite random due to its random initialization of the centroids.

**Credit Card dataset:**

### Credit Card Dataset

| Accuracy | 5% | 10% | 20% | 50% | 80% |
|---|---|---|---|---|---|
| Ensemble | 0.32 | 0.32 | 0.32 | 0.32 | 0.76 |
| Kmeans random | 0.31 | 0.69 | 0.31 | 0.69 | 0.66 |
| Kmeans++ | 0.53 | 0.31 | 0.31 | 0.31 | 0.31 |
| GMM | 0.31 | 0.47 | 0.53 | 0.53 | 0.66 |
| SSL_EM | 0.69 | 0.69 | 0.69 | 0.7 | 0.67 |

Percent of Labeled Data in a batch

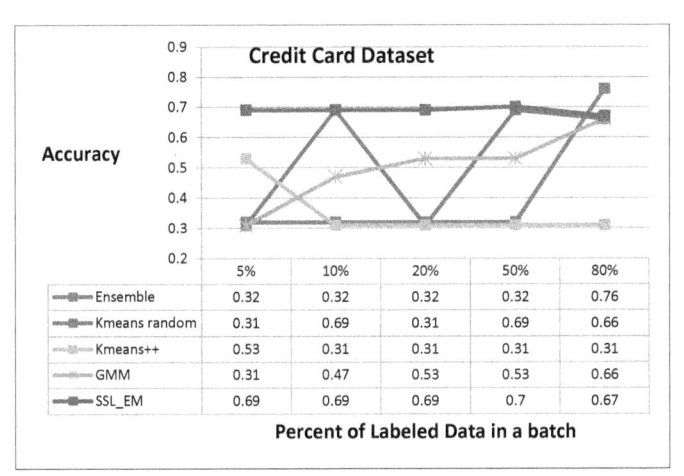

Figure 8-2: Comparison of Accuracy of SSL_EM v/s other algorithms for Credit Card

2. For Credit Card dataset as shown in Figure 8-2, the imbalance is very high. This high imbalance and the available percent of labeled data affects the ensemble classifier, hence its performance is very poor. The imbalance affects the clustering algorithms as well in creating very few minority class so much so that their AUROC values is also very low. For SSL_EM however, the high imbalance and the percent label do not have adverse effect and the performance is consistent.

**Electricity dataset:**

3. For Electricity dataset, the imbalance is very high. This high imbalance and the percent label available affects the ensemble classifier hence its performance is very poor as shown in Figure 8-3 above. SSL_EM and ensemble classifier have almost similar performance for this dataset. K-means++, k-means random and GMM have performance almost similar to each other with very low accuracy.

# Research Findings and Analysis

## Electricity dataset

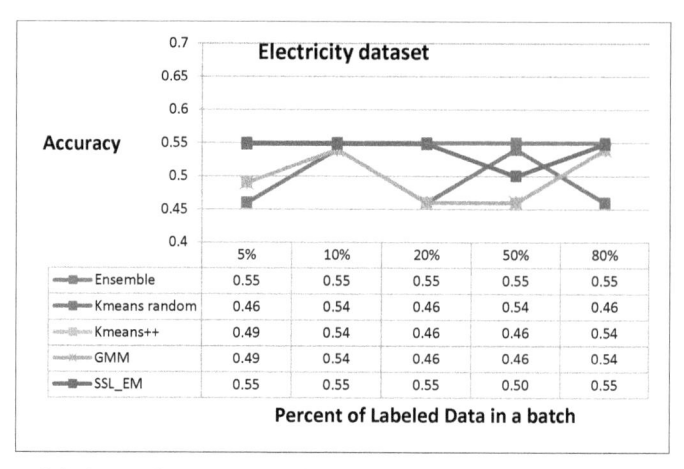

| Percent of Labeled Data in a batch | 5% | 10% | 20% | 50% | 80% |
|---|---|---|---|---|---|
| Ensemble | 0.55 | 0.55 | 0.55 | 0.55 | 0.55 |
| Kmeans random | 0.46 | 0.54 | 0.46 | 0.54 | 0.46 |
| Kmeans++ | 0.49 | 0.54 | 0.46 | 0.46 | 0.54 |
| GMM | 0.49 | 0.54 | 0.46 | 0.46 | 0.54 |
| SSL_EM | 0.55 | 0.55 | 0.55 | 0.50 | 0.55 |

**Figure 8-3: Comparison Accuracy of SSL_EM v/s other algorithms for Electricity dataset**

**Hyperplane datasets:**

4. In Hyperplane datasets, the available labeled data has very little effect on the performance of SSL_EM. For Hyperplane 1 and Hyperplane 2 both, it can be seen in Figure 8-4 (a) and (b) that the values of accuracy are almost consistent for SSL_EM. Both the datasets have higher performance for SSL_EM for all percent labeled under consideration. This means that the SSL_EM can perform very well in the absence of fully labeled data. Ensemble classification, however, has higher performance for 80% labeled due to its supervised nature.

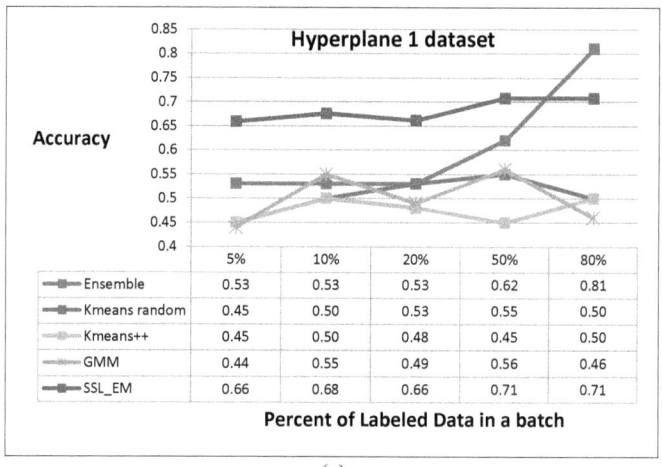

## Hyperplane 1 dataset

| Percent of Labeled Data in a batch | 5% | 10% | 20% | 50% | 80% |
|---|---|---|---|---|---|
| Ensemble | 0.53 | 0.53 | 0.53 | 0.62 | 0.81 |
| Kmeans random | 0.45 | 0.50 | 0.53 | 0.55 | 0.50 |
| Kmeans++ | 0.45 | 0.50 | 0.48 | 0.45 | 0.50 |
| GMM | 0.44 | 0.55 | 0.49 | 0.56 | 0.46 |
| SSL_EM | 0.66 | 0.68 | 0.66 | 0.71 | 0.71 |

(a)

# Research Findings and Analysis

### Hyperplane2 dataset

| | 5% | 10% | 20% | 50% | 80% |
|---|---|---|---|---|---|
| Ensemble | 0.46 | 0.46 | 0.46 | 0.55 | 0.81 |
| Kmeans random | 0.44 | 0.57 | 0.47 | 0.49 | 0.46 |
| Kmeans++ | 0.55 | 0.42 | 0.48 | 0.49 | 0.46 |
| GMM | 0.54 | 0.60 | 0.52 | 0.55 | 0.47 |
| SSL_EM | 0.63 | 0.65 | 0.64 | 0.66 | 0.65 |

Percent of Labeled Data in a batch

(b)

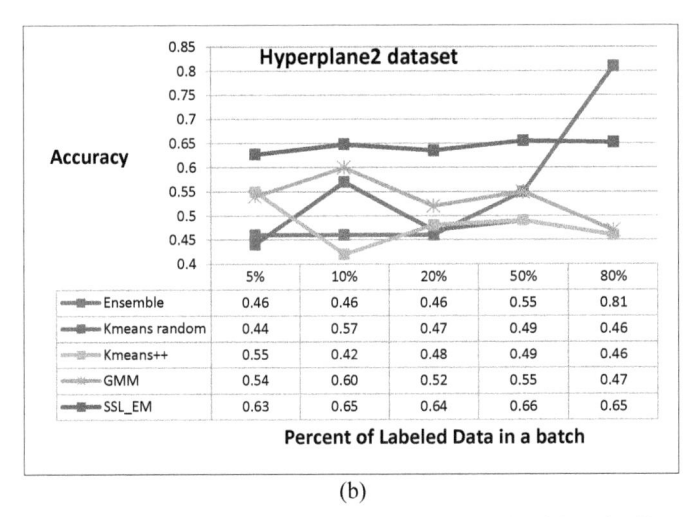

**Figure 8-4: Comparison of Accuracy of SSL_EM v/s other algorithms for Hyperplane datasets**

### SEA datasets:

5. For SEA datasets also the available labeled data has very little effect on the performance of SSL_EM. It can be seen from Figure 8-5 (a) and (b) that the values of accuracy are almost consistent with ensemble classifier for SEA1 but better for SEA2. However, the ensemble classifier performs well with 80% labeled data due to its supervised nature.

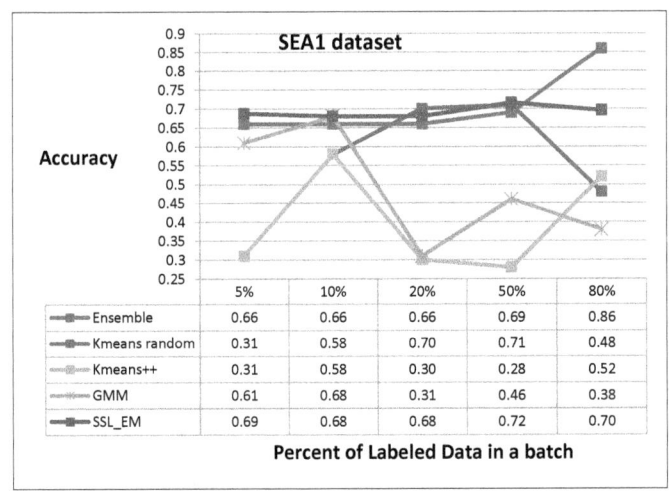

### SEA1 dataset

| | 5% | 10% | 20% | 50% | 80% |
|---|---|---|---|---|---|
| Ensemble | 0.66 | 0.66 | 0.66 | 0.69 | 0.86 |
| Kmeans random | 0.31 | 0.58 | 0.70 | 0.71 | 0.48 |
| Kmeans++ | 0.31 | 0.58 | 0.30 | 0.28 | 0.52 |
| GMM | 0.61 | 0.68 | 0.31 | 0.46 | 0.38 |
| SSL_EM | 0.69 | 0.68 | 0.68 | 0.72 | 0.70 |

Percent of Labeled Data in a batch

(a)

＃ Research Findings and Analysis

**SEA2 dataset**

| | 5% | 10% | 20% | 50% | 80% |
|---|---|---|---|---|---|
| Ensemble | 0.58 | 0.58 | 0.58 | 0.62 | 0.80 |
| Kmeans random | 0.47 | 0.64 | 0.47 | 0.29 | 0.64 |
| Kmeans++ | 0.53 | 0.64 | 0.47 | 0.29 | 0.36 |
| GMM | 0.48 | 0.46 | 0.47 | 0.7 | 0.56 |
| SSL_EM | 0.74 | 0.84 | 0.81 | 0.78 | 0.78 |

Percent of Labeled Data in a batch

(b)

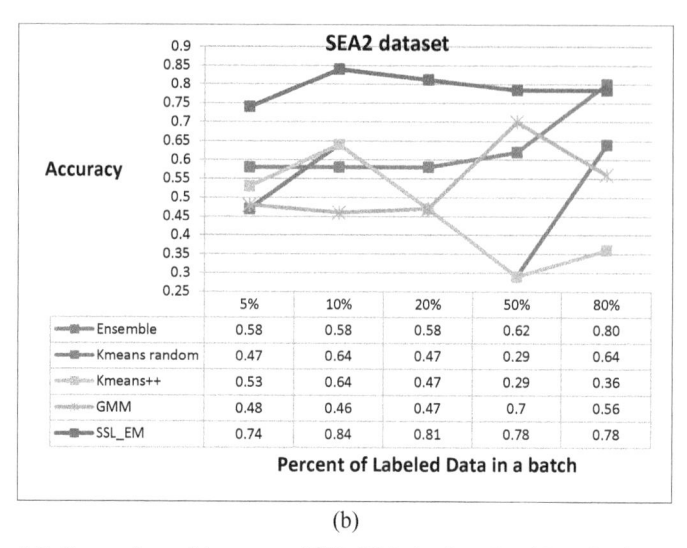

**Figure 8-5: Comparison of Accuracy of SSL_EM v/s other algorithms for SEA datasets**

The available percent of labeled data has no effect on the standard clustering algorithms due to their inability to use the labeled data to their advantage. On the other hand, the labeled data affects the performance of supervised learning ensemble classifiers in the sense that increase in the percent of labeled data increases their performance. As for the proposed semi-supervised clustering technique, SSL_EM, it is able to use the labeled data to its advantage to form labeled clusters. The performance of SSL_EM is consistent and better as compared to other clustering algorithms and the ensemble classifiers. Thus **Hypothesis 1 is true and accepted**.

*8.2.1.2 Test for AUROC values of SSL_EM*

***Hypothesis 2:*** SSL_EM has a higher value for the area under the ROC (AUROC) curve as compared to that of the standard clustering and ensemble classification technique.

**Experimental Setup:**

The experiments are the same as those of previous section. While executing the previous experiment for testing the accuracy values, AUROC values were also evaluated. The AUROC values are the performance measure of classification problem at various thresholds and it tells the classification model's capability to

## Research Findings and Analysis

distinguish the two classes. The AUROC values generally fall in the 0 – 1 range. AUROC values those are higher than 0.5 are desirable for any model. As in the earlier experimentation of section 8.2.1.1, the datasets are assumed to p% labeled and the roc score is calculated for the unseen batch based on the prediction by the model. The roc values for different p% labeled data for their respective methods are plotted and used for comparing and evaluation of SSL_EM.

**Analysis:**

From the results shown in graphs below, the proposed SSL_EM has a better AUROC value as compared to other techniques. The results show that the AUROC values for all datasets except Electricity are above 0.6, which proves that SSL_EM is able to classify minority class instances better than the comparative algorithms.

1. For Spambase dataset and Credit Card dataset as seen in Figure 8-6 and 8-7, the AUROC values are greater than 0.6 and are almost constant across the different percent labeled. Spambase and Credit card both have imbalance ratio higher than the SEA datasets. But SSL_EM is able to form clusters for both minority and majority class and is able to classify minority recall better than the comparative algorithms.

**Spambase dataset:**

### Spambase dataset

| AUROC | 5% | 10% | 20% | 50% | 80% |
|---|---|---|---|---|---|
| Ensemble | 0.50 | 0.50 | 0.50 | 0.50 | 0.61 |
| Kmeans random | 0.50 | 0.50 | 0.50 | 0.50 | 0.50 |
| Kmeans++ | 0.50 | 0.50 | 0.50 | 0.50 | 0.50 |
| GMM | 0.50 | 0.50 | 0.50 | 0.50 | 0.50 |
| SSL_EM | 0.63 | 0.62 | 0.62 | 0.63 | 0.63 |

Percent of Labeled Data in a batch

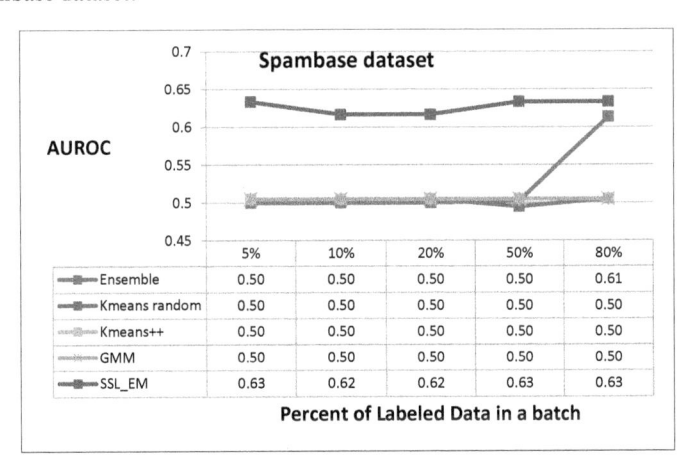

**Figure 8-6: Comparison of AUROC of SSL_EM v/s other algorithms for Spambase datasets**

Research Findings and Analysis

**Credit Card dataset:**

**Credit Card Dataset**

AUROC

| | 5% | 10% | 20% | 50% | 80% |
|---|---|---|---|---|---|
| Ensemble | 0.50 | 0.50 | 0.50 | 0.50 | 0.50 |
| Kmeans random | 0.42 | 0.58 | 0.42 | 0.58 | 0.57 |
| Kmeans++ | 0.55 | 0.42 | 0.42 | 0.42 | 0.43 |
| GMM | 0.42 | 0.45 | 0.55 | 0.55 | 0.57 |
| SSL_EM | 0.58 | 0.58 | 0.58 | 0.61 | 0.59 |

Percent of Labeled Data in a batch

Figure 8-7: Comparison of AUROC of SSL_EM v/s other algorithms for Credit Card datasets

**Electricity dataset:**

**Electricity dataset**

AUROC

| | 5% | 10% | 20% | 50% | 80% |
|---|---|---|---|---|---|
| Ensemble | 0.5 | 0.50 | 0.5 | 0.50 | 0.50 |
| Kmeans random | 0.46 | 0.54 | 0.46 | 0.54 | 0.46 |
| Kmeans++ | 0.46 | 0.54 | 0.46 | 0.46 | 0.54 |
| GMM | 0.45 | 0.50 | 0.50 | 0.46 | 0.50 |
| SSL_EM | 0.50 | 0.50 | 0.5 | 0.47 | 0.50 |

Percent of Labeled Data in a batch

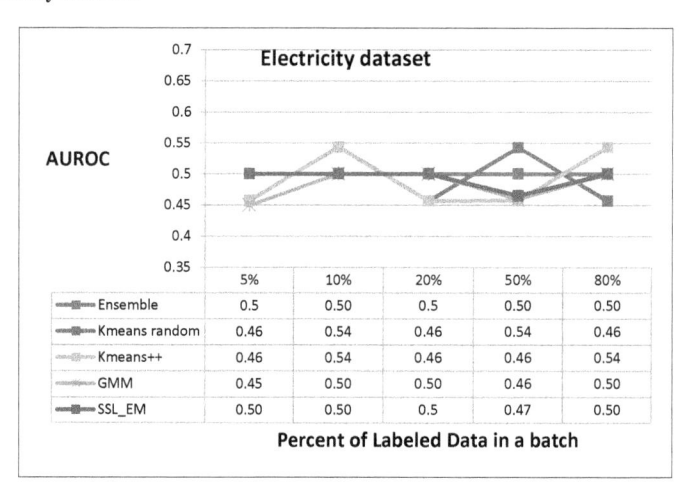

Figure 8-8: Comparison of AUROC of SSL_EM v/s other algorithms for Electricity datasets

2. For Electricity dataset as shown in Figure 8-8, it was observed that the ensemble classification and SSL_EM have the highest value of AUROC but this value 0.5 is indecisive for the model. It is unable to provide a clear distinction for positive and negative classes. For other clustering techniques AUROC value < 0.5 means they have worst separability of classes and are unable to classify the positive

## Research Findings and Analysis

class correctly for most of the percent labeled data. For 10% and 80% k-means++ and k-means random have a better AUROC values but these again are not dependent on the labeled data and are not consistent every time.

**Hyperplane datasets:**

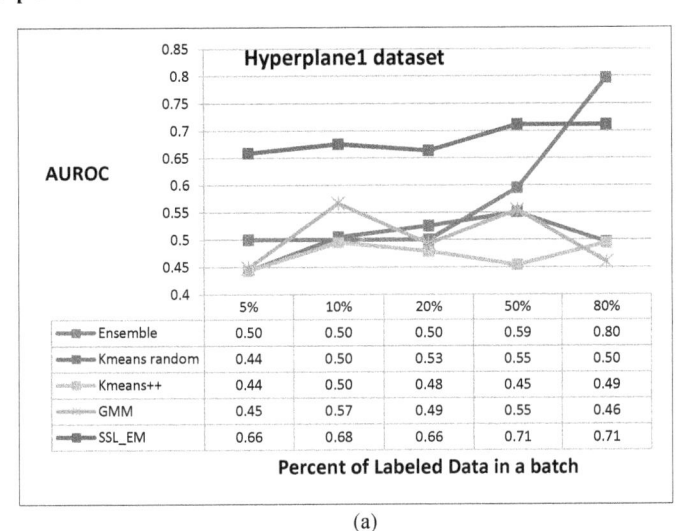

### Hyperplane1 dataset

AUROC

| | 5% | 10% | 20% | 50% | 80% |
|---|---|---|---|---|---|
| Ensemble | 0.50 | 0.50 | 0.50 | 0.59 | 0.80 |
| Kmeans random | 0.44 | 0.50 | 0.53 | 0.55 | 0.50 |
| Kmeans++ | 0.44 | 0.50 | 0.48 | 0.45 | 0.49 |
| GMM | 0.45 | 0.57 | 0.49 | 0.55 | 0.46 |
| SSL_EM | 0.66 | 0.68 | 0.66 | 0.71 | 0.71 |

Percent of Labeled Data in a batch

(a)

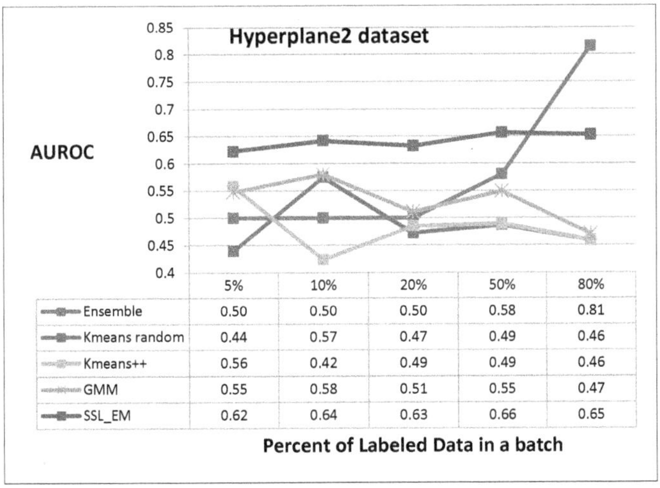

### Hyperplane2 dataset

AUROC

| | 5% | 10% | 20% | 50% | 80% |
|---|---|---|---|---|---|
| Ensemble | 0.50 | 0.50 | 0.50 | 0.58 | 0.81 |
| Kmeans random | 0.44 | 0.57 | 0.47 | 0.49 | 0.46 |
| Kmeans++ | 0.56 | 0.42 | 0.49 | 0.49 | 0.46 |
| GMM | 0.55 | 0.58 | 0.51 | 0.55 | 0.47 |
| SSL_EM | 0.62 | 0.64 | 0.63 | 0.66 | 0.65 |

Percent of Labeled Data in a batch

(b)

**Figure 8-9: Comparison of AUROC of SSL_EM v/s other algorithms for Hyperplane datasets**

Research Findings and Analysis

3. For the Hyperplane datasets, the increase in percent label also increases the ROC score and it is seen to be above 0.6 which is desirable as seen in Figure 8-9. With different percent of labeled data available, the SSL_EM is able to give a consistent performance in terms of the ROC score. The ensemble classifier has better ROC value for 50% and above, whereas the clustering algorithms are inconsistent due to their random initializations.

**SEA datasets:**

**SEA1 dataset**

| AUROC | 5% | 10% | 20% | 50% | 80% |
|---|---|---|---|---|---|
| Ensemble | 0.50 | 0.50 | 0.50 | 0.56 | 0.80 |
| Kmeans random | 0.27 | 0.58 | 0.72 | 0.76 | 0.46 |
| Kmeans++ | 0.27 | 0.58 | 0.28 | 0.24 | 0.54 |
| GMM | 0.47 | 0.71 | 0.41 | 0.48 | 0.36 |
| SSL_EM | 0.73 | 0.73 | 0.73 | 0.75 | 0.74 |

Percent of Labeled Data in a batch

(a)

**SEA2 dataset**

| AUROC | 5% | 10% | 20% | 50% | 80% |
|---|---|---|---|---|---|
| Ensemble | 0.50 | 0.50 | 0.50 | 0.55 | 0.77 |
| Kmeans random | 0.47 | 0.62 | 0.48 | 0.30 | 0.63 |
| Kmeans++ | 0.53 | 0.62 | 0.48 | 0.30 | 0.37 |
| GMM | 0.44 | 0.49 | 0.49 | 0.68 | 0.55 |
| SSL_EM | 0.74 | 0.84 | 0.80 | 0.77 | 0.77 |

Percent of Labeled Data in a batch

(b)

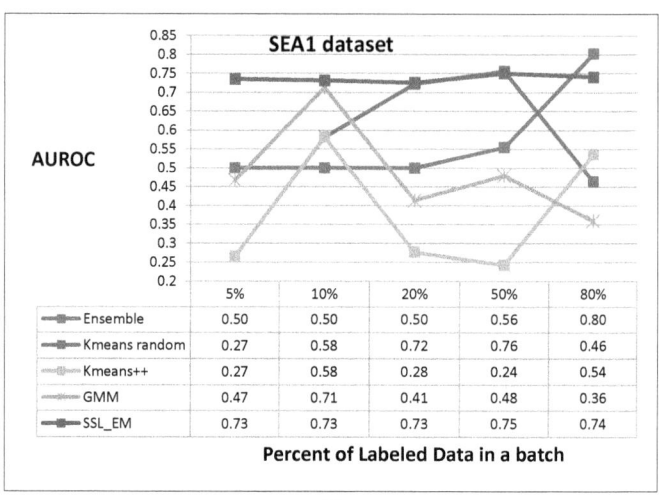

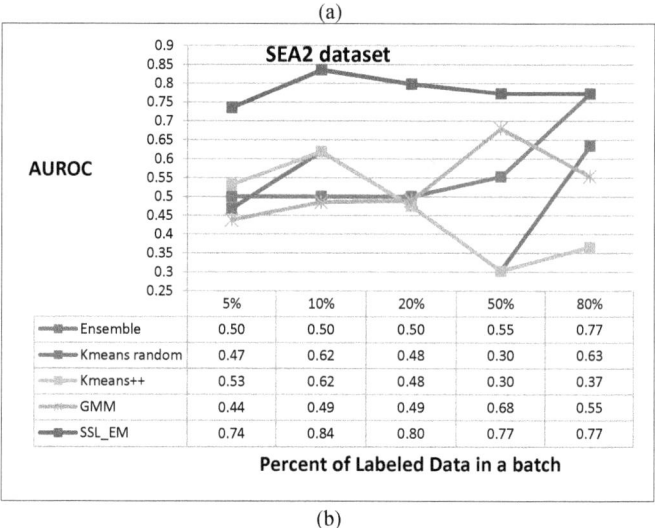

Figure 8-10: Comparison of AUROC of SSL_EM v/s other algorithms for SEA datasets

# Research Findings and Analysis

4. For the SEA datasets, the increase in the percent labeled does not cause variation or increase in the ROC score, the values are more or less constant as shown in Figure 8-10 (a) and (b). SEA dataset has higher imbalance than Hyperplane and the results prove that with higher imbalance also SSL_EM is able to get higher ROC score around 0.7 – 0.8 for both SEA1 and SEA2.

The ROC values for SSL_EM are more than 0.5 for most of the datasets which proves that the proposed SSL_EM have better separability for classes and is able to classify minority class better as compared to other algorithms. Hence the **Hypothesis 2 is true and accepted.**

### 8.2.2 Test Case to assess the cluster purity check module of SSC

***Hypothsis 3:*** To evaluate the cluster purity check module improves the performance of SSC in terms of accuracy.

**Experimental Setup:**

The SSL_EM performs better than the clustering techniques as well as an ensemble classifier. However, the performance can be increased further if the clusters formed by SSL_EM are cohesive. Hence, the cluster purity check and the pure cluster creation modules were added. This test particularly evaluates the recommended cluster purity check module and confirms if it is a valuable addition to the semi-supervised classification model. Each of the clusters created through SSL_EM is checked to see if the clusters formed are cohesive and homogenous. If not, the clusters are split into homogeneous clusters and label propagation is then performed. For this test, the performance of SSC is evaluated with and without the cluster purity check module. The performance of the model without cluster purity check is evaluated in the previous test cases of 8.2.1. The accuracy values from those test cases are considered as base accuracy i.e. accuracy evaluated without adding cluster purity check. Then the accuracy after adding cluster purity check module is evaluated and comparison is made for all the datasets in consideration.

**Results:**

The graphs shown in Figure 8-11 compare the performance of the proposed ensemble model with and without cluster purity check module. The X-axis depicts

# Research Findings and Analysis

the available percent of labeled data in each batch and the Y-axis shows the overall accuracy value for that percent of labeled data.

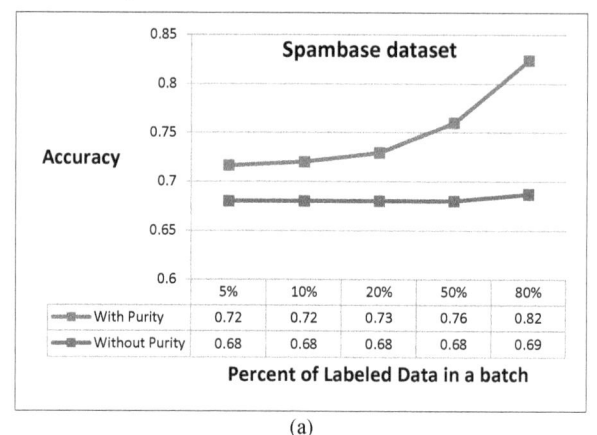

### Spambase dataset

| | 5% | 10% | 20% | 50% | 80% |
|---|---|---|---|---|---|
| With Purity | 0.72 | 0.72 | 0.73 | 0.76 | 0.82 |
| Without Purity | 0.68 | 0.68 | 0.68 | 0.68 | 0.69 |

Percent of Labeled Data in a batch

(a)

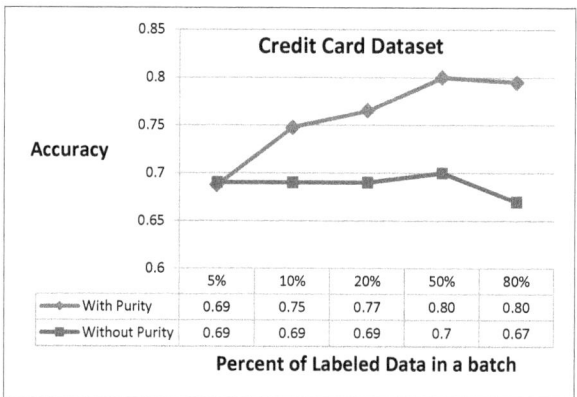

### Credit Card Dataset

| | 5% | 10% | 20% | 50% | 80% |
|---|---|---|---|---|---|
| With Purity | 0.69 | 0.75 | 0.77 | 0.80 | 0.80 |
| Without Purity | 0.69 | 0.69 | 0.69 | 0.7 | 0.67 |

Percent of Labeled Data in a batch

(b)

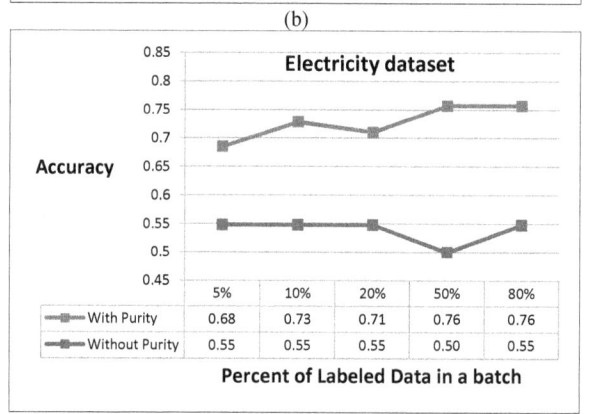

### Electricity dataset

| | 5% | 10% | 20% | 50% | 80% |
|---|---|---|---|---|---|
| With Purity | 0.68 | 0.73 | 0.71 | 0.76 | 0.76 |
| Without Purity | 0.55 | 0.55 | 0.55 | 0.50 | 0.55 |

Percent of Labeled Data in a batch

(c)

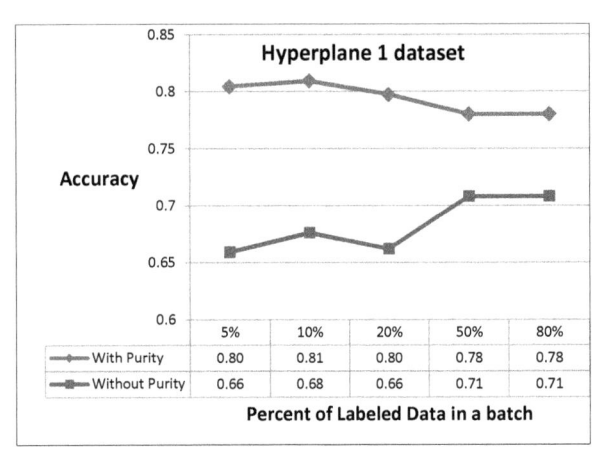

# Research Findings and Analysis

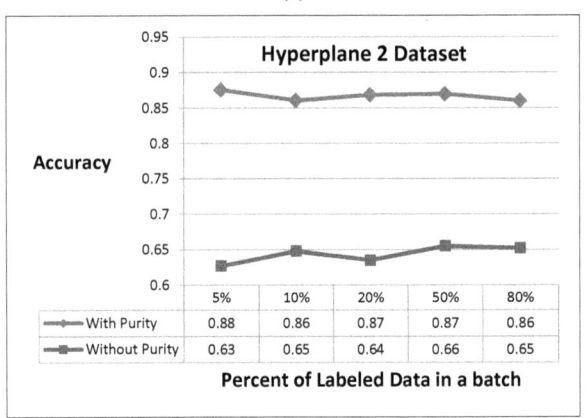

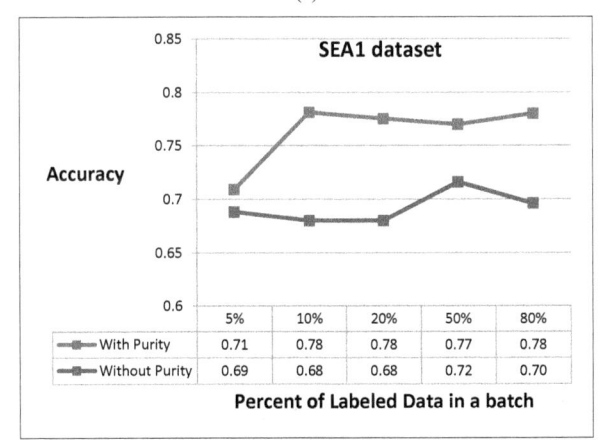

**Hyperplane 1 dataset**

| Percent of Labeled Data in a batch | 5% | 10% | 20% | 50% | 80% |
|---|---|---|---|---|---|
| With Purity | 0.80 | 0.81 | 0.80 | 0.78 | 0.78 |
| Without Purity | 0.66 | 0.68 | 0.66 | 0.71 | 0.71 |

(d)

**Hyperplane 2 Dataset**

| Percent of Labeled Data in a batch | 5% | 10% | 20% | 50% | 80% |
|---|---|---|---|---|---|
| With Purity | 0.88 | 0.86 | 0.87 | 0.87 | 0.86 |
| Without Purity | 0.63 | 0.65 | 0.64 | 0.66 | 0.65 |

(e)

**SEA1 dataset**

| Percent of Labeled Data in a batch | 5% | 10% | 20% | 50% | 80% |
|---|---|---|---|---|---|
| With Purity | 0.71 | 0.78 | 0.78 | 0.77 | 0.78 |
| Without Purity | 0.69 | 0.68 | 0.68 | 0.72 | 0.70 |

(f)

## Research Findings and Analysis

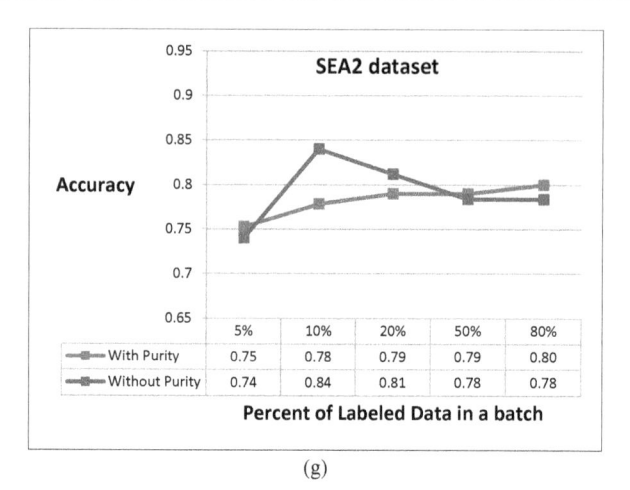

|  | 5% | 10% | 20% | 50% | 80% |
|---|---|---|---|---|---|
| With Purity | 0.75 | 0.78 | 0.79 | 0.79 | 0.80 |
| Without Purity | 0.74 | 0.84 | 0.81 | 0.78 | 0.78 |

SEA2 dataset — Percent of Labeled Data in a batch vs Accuracy

(g)

**Figure 8-11: Comparison of Accuracy with and without cluster purity check module for SSC**

**Analysis:**

It is observed from the graphs of Figure 8-11 (a) to (g) that the cluster purity check module did improve the performance of the semi-supervised clustering based classification module. The splitting of clusters in each model of the ensemble based on their purity have proved beneficial for almost all datasets except SEA2. It is also observed that for imbalanced data sets the increase in the available labeled data amounts to creating better clusters and increase in the accuracy. For balanced data, Hyperplane1 and Hyperplane2 there is an increase in the accuracy after cluster purity check but it is consistent and does not depend on the labeled data. Since more than 90% of the times the cluster purity check module has improved the performance, the **Hypothesis 3 is accepted.**

### 8.2.3 Test Cases for Drift Detection and Adaptation of ASSCCMI

Another main objective of the research is detecting drifts in presence of imbalanced data streams. This test case evaluates the performance of ASSCCMI in detecting drifts present in the imbalanced data streams and adapt to it. To evaluate drift detection and adaptation two test cases are defined. The test case 8.2.3.1 proves that ASSCCMI is able to detect sudden, gradual and incremental drifts and test case 8.2.3.2 is used to prove that drift detection and adaptation module further improves the performance of ASSCCMI. To perform this experiment, drift induced synthetic datasets viz., SEA1, SEA2, Hyperplane1 and Hyperplane2 are generated using the MOA - Massive Online Analysis [112] framework. Different drifts are induced in each of these datasets using drift generators in MOA (Refer Appendix A.1).

### 8.2.3.1 Test for evaluating Drift Detection and Adaptation capability of ASSCCMI on data streams

*Hypothesis 4:* The ASSCCMI is able to detect different types of drifts in imbalanced as well as balanced data streams and adapt to them successfully.

**Experimental Setup:**

ASSCCMI is evaluated for Sudden, Gradual and Incremental drift detection and adaptation on the respective data streams. The graphs shown below depict the ability of ASSCCMI to detect and adapt to drift. The specific instances at which the drifts are detected, the cumulative accuracy and mean square error before and after the drifts as well as subsequent drift detection if any are listed in the below each graphs.

**Analysis:**

ASSCCMI is able to detect sudden, gradual and incremental drifts. The drifts are detected when the difference in the MSE of the current batch and cumulative error in adaptive window is greater than 1.96 times the standard deviation between the current and previous batches. At this point, the drift detection alarm is triggered. Now, the threshold **T** to differentiate between sudden and gradual or incremental drifts is kept as 0.05. So, if the difference in the accuracy between the previous batches and current batch is greater than **T**, then the drift is considered as sudden drift. After the drifts are detected, ASSCCMI adapts to the drifts by adjusting the adaptive window to the batch where drift was detected and retraining the model thereby improving the accuracy.

*Sudden Drift Detection and Adaptation:*

The sudden drifts are induced in SEA2 dataset at $50000^{th}$ instance using the MOA drift generator query. The graph in Figure 8-12 shows the Accuracy of ASSCCMI

Research Findings and Analysis

after classifying each batch of 1000 instances. The graph starts at 7000 instances as the first 6000 instances are used for training the ensemble model.

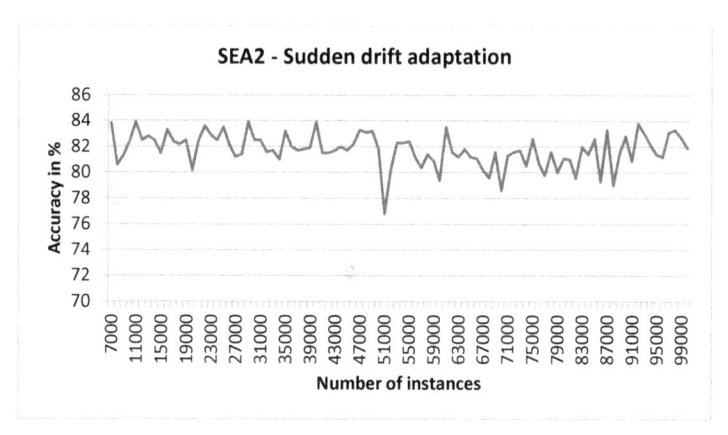

Figure 8-12: Sudden Drift Detection and Adaptation by ASSCCMI

It can be seen from the accuracy graph above, the performance of ASSCCMI drops considerably at 51000$^{th}$ instance as there has been a sudden change in the concept at 50000$^{th}$ instance. This drift is detected and the model is retrained on the window thus making it adapted to the change. Adaptation is evident from the fact that the accuracy of the model increases after 51000$^{th}$ instance. The dataset, the instances at which the drift is detected, and the Accuracy and MSE before the drift instance, at the drift detected instance and after the drift detected instance are shown in Table 8-2 below:

Table 8-2: Analysis of instances with Sudden Drift

| Dataset | Drifts detected at instance | Till the Drift occurs ||| When the Drift occurs ||| After the Drift Adapted |||
|---|---|---|---|---|---|---|---|---|---|---|
| | | MSE | Accuracy % | | MSE | Accuracy % | | MSE | Accuracy % | |
| SEA2 (Sudden Drift) | 51000 | 0.18 | 81 | | 0.23 | 76 | | 0.19 | 81 | |

*Multiple Sudden Drift Detection and Adaptation:*

Multiple sudden drifts were induced in the SEA2 dataset at 25000$^{th}$, 50000$^{th}$ and 75000$^{th}$ instances using the MOA query. The graph shown below in Figure 8-13 depicts the multiple sudden drift detection and adaptation of the ASSCCMI at the respective instances.

**SEA2 - Multiple Sudden drifts adaptation**

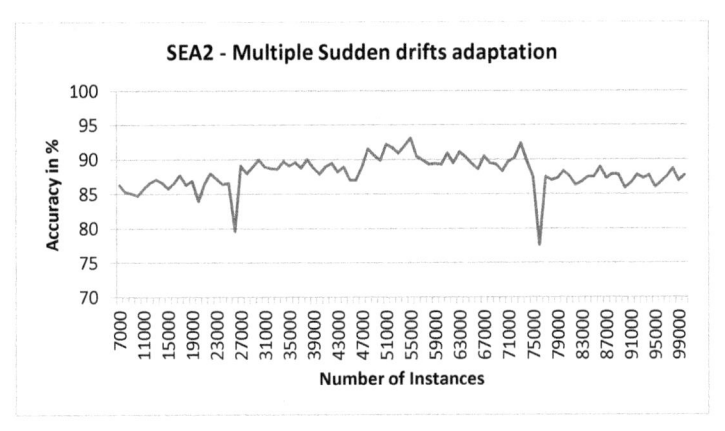

Figure 8-13: Multiple Sudden Drifts Detection and Adaptation by ASSCCMI

There is seen a steep drop in the accuracy at $26000^{th}$ and $76000^{th}$ instance as the drifts were detected at the window between 25000-26000 and 75000-76000 instances. After that the model adapts to the drift and the accuracy is improved. The third drift is detected at the $56000^{th}$ instance instead of $51000^{th}$ as the model accuracy was quite high at 51000 and the MSE difference also did not satisfy the statistical test criteria. However, at $56000^{th}$, the model detects the drift and adapts to it keeping the accuracy high thereafter.

The dataset, the instances at which the drift is detected, and the Accuracy and MSE before the drift instance, at the drift detected instance and after the drift detected instance are shown in Table 8-3 below:

Table 8-3: Analysis of instances with Multiple Sudden Drifts

| Dataset | Drifts detected at instance | Till the Drift occurs MSE | Till the Drift occurs Accuracy % | When the Drift occurs MSE | When the Drift occurs Accuracy % | After the Drift Adapted MSE | After the Drift Adapted Accuracy % |
|---|---|---|---|---|---|---|---|
| SEA2 (Multiple Sudden Drift) | 26000 | 0.13 | 86.6 | 0.20 | 79.6 | 0.11 | 88 |
| | 56000 | 0.07 | 93.1 | 0.09 | 90.5 | 0.10 | 91 |
| | 76000 | 0.13 | 87.5 | 0.23 | 77.5 | 0.13 | 87.5 |

Research Findings and Analysis

## *Gradual Drift Detection and Adaptation:*

The SEA1 dataset has been induced with the gradual drift using the MOA query. The gradual drift is induced in a window of 20000 instances with the centre of the window at 50000$^{th}$ instance. The accuracy of the classifier drops gradually within this window as is visible in graph below. The drift detection module detects a rise in the error of classification from 38000 first where it adapts however, the continuous change in the concept increases the error in accuracy that is later detected at 58000$^{th}$ instance. The graph in Figure 8-14 shows the Accuracy of ASSCCMI after classifying each batch of 1000 instances.

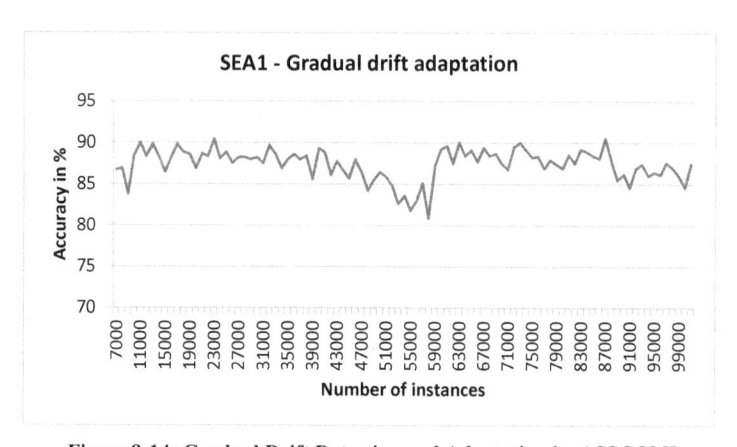

**Figure 8-14: Gradual Drift Detection and Adaptation by ASSCCMI**

It can be seen that the performance of ASSCCMI gradually drops from around 40000$^{th}$ instances. The dataset, the instances at which the drift is detected, and the Accuracy and MSE before the drift instance, at the drift detected instance and after the drift detected instance are shown in Table 8-4 below:

**Table 8-4: Analysis of instances with Gradual Drifts**

|  | Drifts detected at instance | Till the Drift occurs | | When the Drift occurs | | After the Drift Adapted | |
|---|---|---|---|---|---|---|---|
|  |  | MSE | Accuracy % | MSE | Accuracy % | MSE | Accuracy % |
| SEA1 (Gradual Drift) | 38000 | 0.12 | 88 | 0.14 | 85 | 0.11 | 89 |
|  | 58000 | 0.15 | 85 | 0.19 | 81 | 0.13 | 87 |

Research Findings and Analysis

The classifier maintains the accuracy by adapting to the drifts whenever detected. After 60000 instances the accuracy is almost stable thus proving the drift detection and adaptation capability of ASSCCMI for gradual drifts.

*Incremental Drift Detection and Adaptation:*

The incremental drifts are induced by the Hyperplane generator of the MOA. As mentioned in the dataset earlier in section 7.2.5, the incremental drifts are generated by changing magnitude of weights by a factor of 0.01. The graph shown in the Figure 8-15 below shows the accuracy of the classifier at each window of 1000 instances.

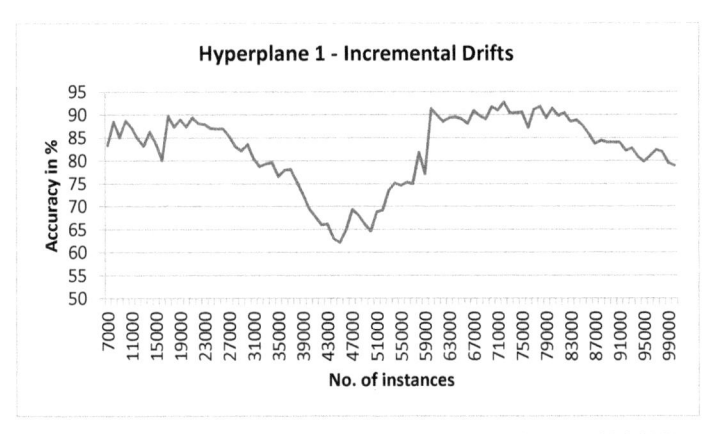

Figure 8-15: Incremental Drift Detection and Adaptation by ASSCCMI

The dataset, the instances at which the drift is detected, and the Accuracy and MSE before the drift instance, at the drift detected instance and after the drift detected instance are shown in Table 8-5 below:

Table 8-5: Analysis of instances with Incremental Drifts

|  | Drifts detected at instance | Till the Drift occurs | | When the Drift occurs | | After the Drift Adapted | |
|---|---|---|---|---|---|---|---|
|  |  | MSE | Accuracy % | MSE | Accuracy % | MSE | Accuracy % |
| Hyperplane1 (Incremental Drifts) | 9000 | 0.11 | 88 | 0.15 | 85 | 0.11 | 88 |
|  | 17000 | 0.16 | 84 | 0.20 | 80 | 0.10 | 90 |

143

# Research Findings and Analysis

| | 59000 | 0.18 | 82 | 0.23 | 77 | 0.09 | 91 |
|---|---|---|---|---|---|---|---|
| | 76000 | 0.09 | 91 | 0.12 | 87 | 0.09 | 91 |

Since the drifts are incremental, the change in the concept happens over a considerable period of time. At $9000^{th}$ instance the first change in concept was detected by the classifier for which it adapts. Since the drift is happening at every window, subsequent drifts are detected at $17000^{th}$, $59000^{th}$ and $76000^{th}$ instance. It can be seen that at these locations the error in classification compared to the cumulative error up till that window is high. This triggered the drift detection and adaptation is performed on previous windows. The ASSCCMI retrains on the previous two windows adjusted by the ADWIN of the module and maintains the classification accuracy till further change detection.

### *Subtle Incremental Drift Detection and Adaptation:*

The subtle incremental drifts are induced by the Hyperplane generator of the MOA. As mentioned in the dataset earlier in section 7.2.5, the incremental drifts are generated by changing magnitude of weights by a factor of 0.001. These introduce the change in the concept very slowly.

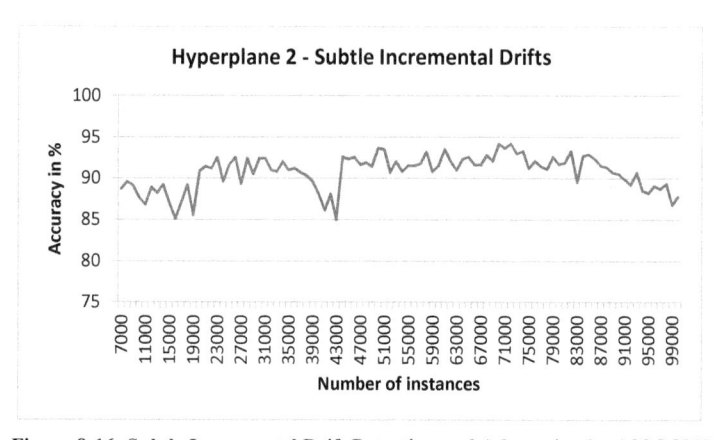

**Figure 8-16: Subtle Incremental Drift Detection and Adaptation by ASSCCMI**

The graph shown in Figure 8-16 shows that whenever a change is detected the ASSCCMI tries to adapt to it. In subtle incremental the concept will change from one concept to another in a very slow manner gradually moving back and forth within the concept.

Research Findings and Analysis

The dataset, the instances at which the drift is detected, and the Accuracy and MSE before the drift instance, at the drift detected instance and after the drift detected instance are shown in Table 8-6 below. The drifts are detected at instances 19000, 24000, 43000, 52000, 59000 and 83000 when the concept has changed such that ASSCCMI is not able to classify correctly. The drift detection module triggers the adaptation module when the error conditions are satisfied i.e. the error in the current window is more than the cumulative error.

Table 8-6: Analysis of instances with Subtle Incremental Drifts

|  | Drifts detected at instance | Till the Drift occurs ||  When the Drift occurs || After the Drift Adapted ||
|---|---|---|---|---|---|---|---|
|  |  | MSE | Accuracy % | MSE | Accuracy % | MSE | Accuracy % |
| Hyperplane2 (Subtle Incremental Drifts) | 19000 | 0.11 | 89 | 0.15 | 85 | 0.09 | 90 |
|  | 24000 | 0.08 | 93 | 0.1 | 89 | 0.08 | 92 |
|  | 43000 | 0.12 | 88 | 0.15 | 85 | 0.07 | 92 |
|  | 52000 | 0.07 | 93 | 0.09 | 90 | 0.08 | 92 |
|  | 59000 | 0.07 | 93 | 0.09 | 90 | 0.08 | 92 |
|  | 83000 | 0.07 | 93 | 0.11 | 90 | 0.07 | 93 |

*Mixed Drift Detection and Adaptation:*

Mixed drifts were induced in the SEA2 dataset at various locations of instances using the MOA query. Different types of drifts were induced in the same data stream to test the ability of ASSCCMI in detecting the drifts and adapting to it.

The graph shown below in Figure 8-17 depicts the multiple drift adaptation with 2 gradual drifts at 25000th and 75000th instance with a window of 10000 instances and 1 sudden at 50000th instance.

Research Findings and Analysis

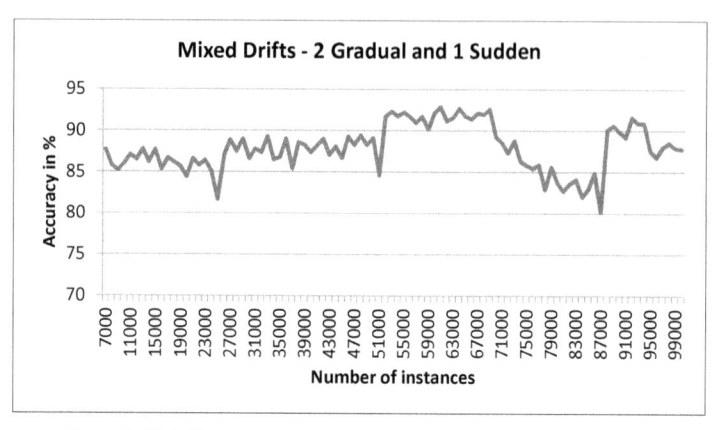

**Figure 8-17: Mixed Drifts Detection and Adaptation by ASSCCMI**

The dataset, the instances at which the drift is detected, and the Accuracy and MSE before the drift instance, at the drift detected instance and after the drift detected instance are shown in Table 8-7 below.

Table 8-7: Analysis of instances with Mixed Drifts

| Mixed Drifts | Drifts detected at instance | Till the Drift occurs MSE | Till the Drift occurs Accuracy % | When the Drift occurs MSE | When the Drift occurs Accuracy % | After the Drift Adapted MSE | After the Drift Adapted Accuracy % |
|---|---|---|---|---|---|---|---|
| Gradual | 25000 | 0.15 | 85.1 | 0.18 | 81.7 | 0.13 | 87 |
| | 37000 | 0.11 | 89 | 0.15 | 85.4 | 0.11 | 88.6 |
| Sudden | 51000 | 0.11 | 89.1 | 0.15 | 84.6 | 0.08 | 91.7 |
| Gradual | 70000 | 0.07 | 92.6 | 0.11 | 89.3 | 0.12 | 88.5 |
| | 87000 | 0.15 | 84.9 | 0.2 | 80.2 | 0.10 | 90.1 |

The gradual drift induced at $25000^{th}$ instance with a window of 20000 instances is detected with first detection at 25000 and the second detection at $37000^{th}$ instance. Later the sudden drift induced at $50000^{th}$ instance is detected at $51000^{th}$ and is adapted by retraining. The second gradual drift induced at $75000^{th}$ with a window of 10000 instances is detected first at 70000 as the accuracy of the classifier drops there and later at 87000 after which the model is retrained to adapt to the drift. The drift detection module triggers the adaptation module when the error conditions are satisfied i.e. the error in the current window is more than the cumulative error.

In all the different types of drifts discussed in this section, the increase in accuracy and reduction in MSE after the drift detection proves the adaption of the model to the concept drift. Thus, it can be concluded that ASSCCMI is able to detect drifts in imbalanced as well as balanced data streams and the **Hypothesis 4 is accepted**.

### 8.2.3.2 Test for assessing improvement in accuracy through Drift Detection and Adaptation of ASSCCMI

*Hypothesis 5:* The Drift Detection and Adaptation module is able to improve the performance of ASSCCMI in terms of its accuracy for scarcely labeled data.

**Experimental Setup:**

The earlier test case proved that ASSCCMI is able to detect different types of drifts in different data streams. In the earlier test case, the focus was on detecting the drift at proper location and the performance of ASSCCMI just before and after the drift is detected. This test case focuses on the overall accuracy of ASSCCMI for available percent of labeled data. This test case is evaluated by comparing the accuracy of ASSCCMI drift detection - adaptation module and without drift detection - adaptation module. The data stream is assumed to be p% labeled in each batch. The test case evaluates the results for p% labeled data.

**Results:**

The graphs in Figure 8-18 (a) to (e) depict the overall accuracy of ASSCCMI with Concept Drift (with CD) module and without Concept Drift (without CD) module for different percent of labeled data available in the batch. The graphs clearly depict that the performance of ASSCCMI improves considerably with the addition of the drift detection and adaptation module.

ASSCCMI is successfully able to detect the drift in the data and adapt to the drift by retraining the model. The newly trained model is added to the ensemble. This enables the model to maintain good classification accuracy even in the case of drifts.

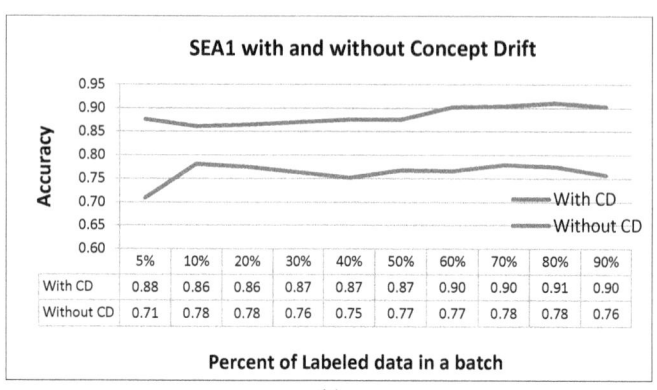

# Research Findings and Analysis

**SEA1 with and without Concept Drift**

| Percent of Labeled data in a batch | 5% | 10% | 20% | 30% | 40% | 50% | 60% | 70% | 80% | 90% |
|---|---|---|---|---|---|---|---|---|---|---|
| With CD | 0.88 | 0.86 | 0.86 | 0.87 | 0.87 | 0.87 | 0.90 | 0.90 | 0.91 | 0.90 |
| Without CD | 0.71 | 0.78 | 0.78 | 0.76 | 0.75 | 0.77 | 0.77 | 0.78 | 0.78 | 0.76 |

(a)

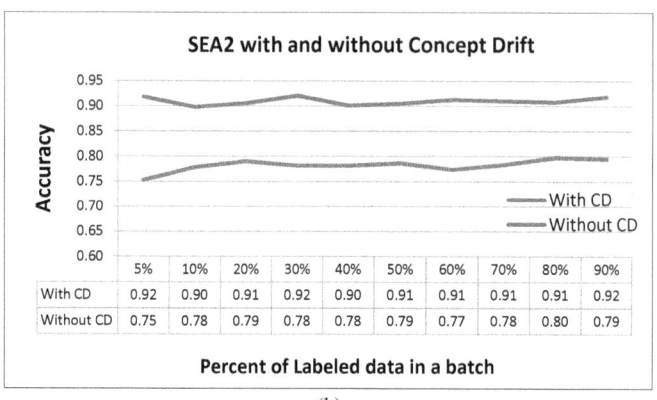

**SEA2 with and without Concept Drift**

| Percent of Labeled data in a batch | 5% | 10% | 20% | 30% | 40% | 50% | 60% | 70% | 80% | 90% |
|---|---|---|---|---|---|---|---|---|---|---|
| With CD | 0.92 | 0.90 | 0.91 | 0.92 | 0.90 | 0.91 | 0.91 | 0.91 | 0.91 | 0.92 |
| Without CD | 0.75 | 0.78 | 0.79 | 0.78 | 0.78 | 0.79 | 0.77 | 0.78 | 0.80 | 0.79 |

(b)

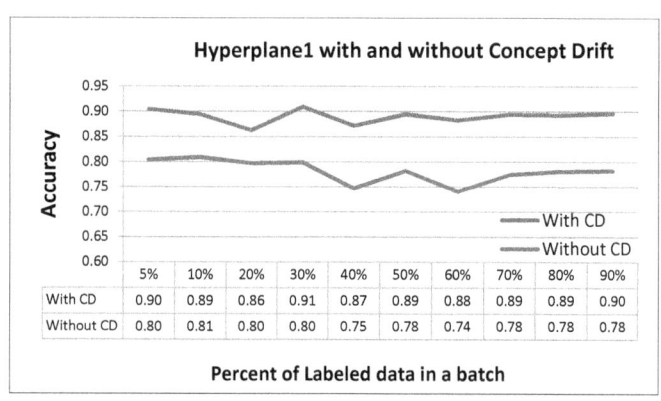

**Hyperplane1 with and without Concept Drift**

| Percent of Labeled data in a batch | 5% | 10% | 20% | 30% | 40% | 50% | 60% | 70% | 80% | 90% |
|---|---|---|---|---|---|---|---|---|---|---|
| With CD | 0.90 | 0.89 | 0.86 | 0.91 | 0.87 | 0.89 | 0.88 | 0.89 | 0.89 | 0.90 |
| Without CD | 0.80 | 0.81 | 0.80 | 0.80 | 0.75 | 0.78 | 0.74 | 0.78 | 0.78 | 0.78 |

(c)

Research Findings and Analysis

### Hyperplane2 with and without Concept Drift

| | 5% | 10% | 20% | 30% | 40% | 50% | 60% | 70% | 80% | 90% |
|---|---|---|---|---|---|---|---|---|---|---|
| With CD | 0.89 | 0.90 | 0.90 | 0.91 | 0.89 | 0.90 | 0.89 | 0.91 | 0.90 | 0.89 |
| Without CD | 0.88 | 0.86 | 0.87 | 0.87 | 0.86 | 0.87 | 0.88 | 0.88 | 0.86 | 0.85 |

Percent of Labeled data in a batch

(d)

### Electricity with and without Concept Drift

| | 5% | 10% | 20% | 30% | 40% | 50% | 60% | 70% | 80% | 90% |
|---|---|---|---|---|---|---|---|---|---|---|
| With CD | 0.79 | 0.77 | 0.79 | 0.77 | 0.78 | 0.77 | 0.77 | 0.81 | 0.79 | 0.79 |
| Without CD | 0.68 | 0.73 | 0.71 | 0.74 | 0.74 | 0.76 | 0.76 | 0.74 | 0.75 | 0.78 |

Percent of Labeled data in a batch

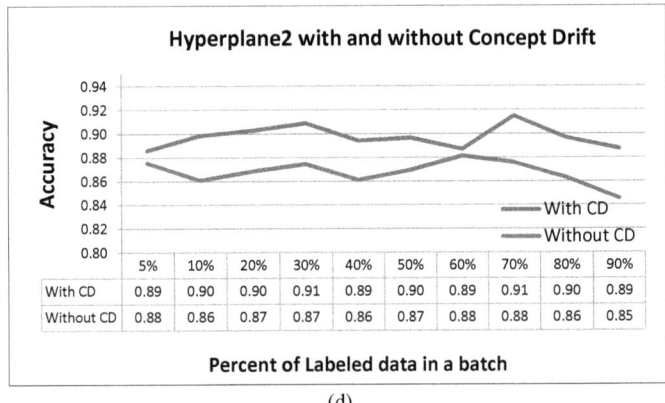

(e)

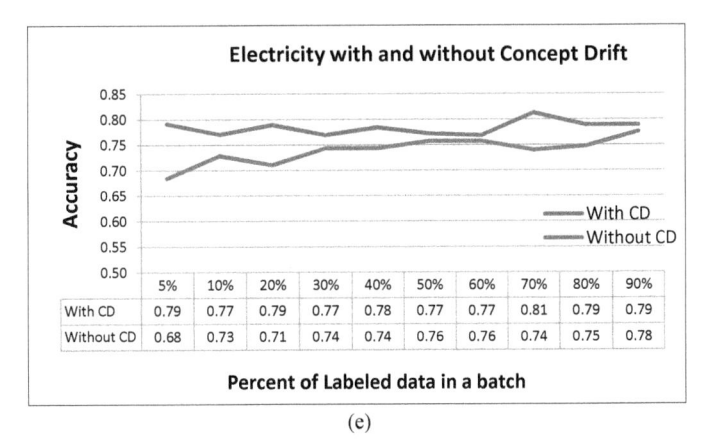

Figure 8-18: Performance of ASSCCMI with and without Drift Detection and Adaptation

The overall improvement in the accuracy of the model due to drift adaptation is shown in Table 8-8.

Table 8-8: Comparison of Average Accuracy with and without Drift Detection and Adaptation

| | SEA1 | SEA2 | Hyperplane1 | Hyperplane2 | Electricity |
|---|---|---|---|---|---|
| Accuracy with CD | 89% ± 2% | 91% ± 2% | 89% ± 2% | 90% ± 1% | 79% ± 2% |
| Accuracy without CD | 74% ± 4% | 77% ± 3% | 77% ± 3% | 87% ± 1% | 73% ± 5% |

# Research Findings and Analysis

**Analysis:**

The average accuracy in the datasets with drifts sees a considerable increase in the overall accuracy percentage. It should be noted that the datasets are balanced (Hyperplane1 & Hyperplane2) as well as imbalanced (Electricity, SEA1 & SEA2). It can be seen that the imbalance in data does not affect the performance of ASSCCMI. This proves that ASSCCMI is able to perform drift detection even in the presence of imbalanced data. The overall increase in average accuracy due to drift detection and adaptation for each of the dataset is as shown in Table 8-9 below:

**Table 8-9: Percentage increase in Average Accuracy of ASSCCMI due to Drift Adaptation**

|  | SEA1 | SEA2 | Hyperplane1 | Hyperplane2 | Electricity |
|---|---|---|---|---|---|
| Increase in Accuracy due to Concept Drift Adaptation | 15% ± 3% | 14% ± 2% | 12% ± 2% | 3% ± 1% | 6% ± 4% |

Thus, it can be concluded that ASSCCMI is able to detect and adapt to concept drift even in scarcely labeled data. Hence, the **Hypothesis 5 is true and accepted.**

## 8.2.4 Test Case for Overall Accuracy of ASSCCMI

The test cases designed and evaluated in sections 8.2.1 to 8.2.3 were used to assess the performance of different modules of ASSCCMI viz., the proposed semi-supervised clustering approach, the recommended cluster purity check and the proposed drift detection and adaptation. This test case evaluates the main objective of this work i.e. improvising learning in imbalanced data streams with scarcely labeled data. The tests are performed on all the datasets discussed earlier in section 7.2. The test cases are designed to evaluate the performance of ASSCCMI on imbalance data streams, balanced data streams with concept drift and imbalanced data streams with concept drifts. The experiments performed also consider that all the datasets are partially labeled with p% labeled data where p ranges from 5% to 90%. Thus, these test cases evaluate the working of the proposed framework ASSCCMI in absence of fully labeled imbalanced non-stationary data streams.

Research Findings and Analysis

### 8.2.4.1 Test for Accuracy of ASSCCMI on imbalanced data streams

*Hypothesis 6:* The performance of ASSCCMI in terms of overall accuracy of imbalanced data streams is better as compared to the performance of KNNADWIN, OzaBaggingADWIN and SPASC techniques on the same data streams.

**Experimental Setup:**

The ASSCCMI is evaluated on data streams with imbalanced data for this test case. The parameter settings are done as discussed in section 7.5.1. The data stream is assumed to be p% labeled in each batch and for each p% labeled stream, the performance of the model is evaluated. The similar settings like the batch of data, percent of labeled data available and the number of models in the ensemble are maintained for the comparing algorithms. The comparing algorithms, KNNADWIN and OzaBaggingADWIN, are one of the standard stream classification algorithms whose working is closest to ASSCCMI (Refer section 7.4.1). Since ASSCCMI is a semi-supervised learning framework an existing semi-supervised learning framework SPASC from literature is also used for comparison (Refer section 7.4.2). Each technique is executed on all 7 datasets for 5 trials and each trial is executed for 5 to 90% labeled data i.e. total of 50 trials per dataset per technique.

Similar experiments are performed for test cases described in 8.2.4.2 and 8.2.4.3 as well as for test cases of section 8.2.5.

**Results:**

The results of the experiment are plotted in graph shown in Figure 8-19 to 8-21 below. The average accuracy for each percent of available labeled data is plotted for all the algorithms for comparison. The accuracy values are in the range of [0 – 1]. The results show considerable increase in the performance of ASSCCMI in terms of average accuracy as the metric. It should be noted that the imbalance ratio of Electricity, Spambase and CreditCard dataset is 1.36, 1.54 and 2.33 respectively. So CreditCard is the highest imbalanced of all three.

Research Findings and Analysis

**Analysis:** For each dataset, the overall imbalance ratio and the ratio in each batch of data is different. ASSCCMI is unaware of the ratio of distribution. It processes the data as it is seen.

**Spambase Dataset:**

| | 5 | 10 | 20 | 30 | 40 | 50 | 60 | 70 | 80 | 90 |
|---|---|---|---|---|---|---|---|---|---|---|
| ASSCCMI | 0.84 | 0.87 | 0.90 | 0.92 | 0.91 | 0.89 | 0.89 | 0.89 | 0.88 | 0.89 |
| SPASC | 0.64 | 0.63 | 0.62 | 0.62 | 0.62 | 0.62 | 0.61 | 0.62 | 0.62 | 0.61 |
| KNNADWIN | 0.68 | 0.70 | 0.72 | 0.72 | 0.71 | 0.72 | 0.71 | 0.72 | 0.72 | 0.74 |
| OZABAGGINGADWIN | 0.67 | 0.69 | 0.71 | 0.74 | 0.76 | 0.76 | 0.76 | 0.78 | 0.78 | 0.79 |

Percent of Labeled data in a batch

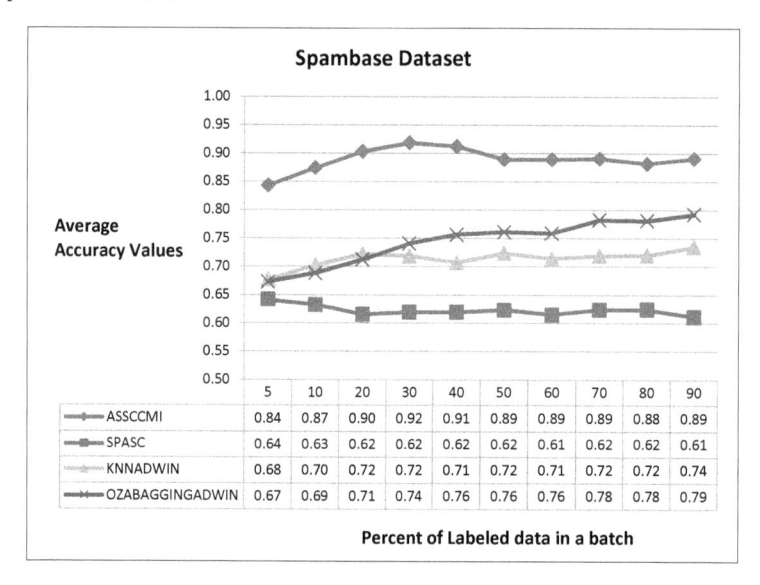

Figure 8-19: Comparison of Average Accuracy of ASSCCMI v/s other algorithms for Spambase dataset

1. For Spambase dataset, the accuracy is considerably high as compared to the other algorithms. ASSCCMI is able to perform better classification than other standard algorithms because of the cluster and label approach of the model. The micro clusters created by this approach are able to retain the information of each class irrespective of its ratio, that aid in improved classification. The performance of ASSCCMI is slightly higher for 30% and 40% data. It is observed that KNNADWIN also behaves similar to ASSCCMI. OzaBaggingADWIN being a supervised classifier that totally depends on the available percent labeled data, its performance is directly proportional to the percent of labeled data. ASSCCMI and SPASC are able to leverage the unlabeled data due to their semi-supervised learning technique. However, SPASC gets affected by the minority ratio. Due to its prequential learning strategy, some of the models in the ensemble do not have

# Research Findings and Analysis

any minority representation. As a result, it is unable to classify the minority class which causes the reduced accuracy.

**Credit Card Dataset:**

**Credit Card Dataset**

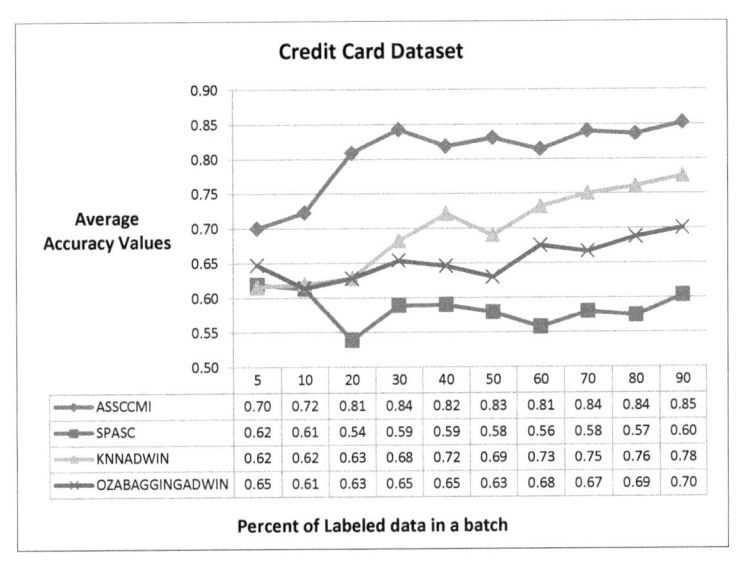

| Percent of Labeled data in a batch | 5 | 10 | 20 | 30 | 40 | 50 | 60 | 70 | 80 | 90 |
|---|---|---|---|---|---|---|---|---|---|---|
| ASSCCMI | 0.70 | 0.72 | 0.81 | 0.84 | 0.82 | 0.83 | 0.81 | 0.84 | 0.84 | 0.85 |
| SPASC | 0.62 | 0.61 | 0.54 | 0.59 | 0.59 | 0.58 | 0.56 | 0.58 | 0.57 | 0.60 |
| KNNADWIN | 0.62 | 0.62 | 0.63 | 0.68 | 0.72 | 0.69 | 0.73 | 0.75 | 0.76 | 0.78 |
| OZABAGGINGADWIN | 0.65 | 0.61 | 0.63 | 0.65 | 0.65 | 0.63 | 0.68 | 0.67 | 0.69 | 0.70 |

**Figure 8-20: Comparison of Average Accuracy of ASSCCMI v/s other algorithms for Credit Card dataset**

2. For CreditCard dataset, the accuracy increases with increase in available labeled data. The average accuracy for ASSCCMI is quite higher than the comparative algorithms. One of the reasons is that, the Credit Card dataset is highly imbalanced in nature. The second reason is that the dataset is small. As a result, each batch size is smaller and with lower percent of labeled data, the available labeled data are very few; sometimes even as few as a single digit in count. Due to such batch formation, the comparative algorithms are affected to a greater extent than the ASSCCMI. Though SPASC is also a semi-supervised approach, it could not achieve better accuracy because of the way it does clustering. It creates cluster on the batch of data it sees. If there are no minority instances in a batch, clusters for that class will not be formed. And in Credit Card, with high imbalance ratio, small batch size and lower number of labeled data; the probability of seeing a minority class instance is very low, sometimes even zero. Hence, SPASC suffers majorly for this dataset. KNNADWIN and

# Research Findings and Analysis

OzaBaggingADWIN also suffer due to this less number of labeled data. As they both train on labeled data, the number of samples available for their training are very few and with high imbalance, they suffer badly for minority class. As a result, the overall accuracy for these algorithms is also poor.

**Electricity Dataset:**

### Electricity Dataset

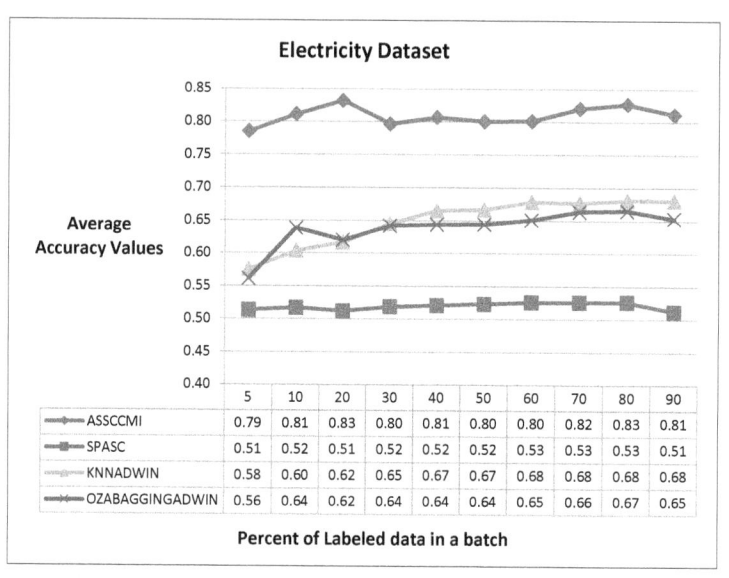

| Percent of Labeled data in a batch | 5 | 10 | 20 | 30 | 40 | 50 | 60 | 70 | 80 | 90 |
|---|---|---|---|---|---|---|---|---|---|---|
| ASSCCMI | 0.79 | 0.81 | 0.83 | 0.80 | 0.81 | 0.80 | 0.80 | 0.82 | 0.83 | 0.81 |
| SPASC | 0.51 | 0.52 | 0.51 | 0.52 | 0.52 | 0.52 | 0.53 | 0.53 | 0.53 | 0.51 |
| KNNADWIN | 0.58 | 0.60 | 0.62 | 0.65 | 0.67 | 0.67 | 0.68 | 0.68 | 0.68 | 0.68 |
| OZABAGGINGADWIN | 0.56 | 0.64 | 0.62 | 0.64 | 0.64 | 0.64 | 0.65 | 0.66 | 0.67 | 0.65 |

**Figure 8-21: Comparison of Average Accuracy of ASSCCMI v/s other algorithms for Electricity dataset**

3. For Electricity dataset, the average accuracy of ASSCCMI is quite higher than other algorithms. This dataset has lower imbalance than Spambase and Credit Card datasets. Apart from being imbalanced, this dataset is also known to have concept drift. However, the type of drifts and the instances where the drifts occurred are unknown; hence this dataset is included in imbalanced test case and discussed here. The ASSCCMI, however, is able to detect drift and adapt to these implicit drifts, which leads to the increase in the accuracy of the model as compared to other algorithms. The average accuracy for all percent labeled data is about 80%.

The dataset is observed to have long consecutive periods of DOWNs (majority class) followed by long periods of UPs (minority class). But, the cluster and label

# Research Findings and Analysis

approach of ASSCCMI helps to retain the minority class information in the micro cluster. And this helps for better classification accuracy. However, this same characteristic affects SPASC, KNNADWIN and OzaBaggingADWIN adversely. These algorithms do not get to see the minority class information for a long time. As a result they show poor performance in terms of accuracy. It will also be seen that their minority class recall for these algorithms are very poor as compared to ASSCCMI.

The average accuracy of each of the algorithms for the datasets is shown in Table 8-10 below:

**Table 8-10: Comparison of Average Accuracy values for Imbalanced data streams**

|  | Spambase | CreditCard | Electricity |
| --- | --- | --- | --- |
| ASSCCMI | 88% ± 4% | 78% ± 8% | 80% ± 2% |
| SPASC | 62% ± 2% | 58% ± 4% | 52% ± 1% |
| KNNADWIN | 71% ± 3% | 70% ± 8% | 63% ± 5% |
| OzaBaggingADWIN | 73% ± 6% | 65% ± 5% | 62% ± 6% |

Thus, it can be seen that for imbalanced data streams ASSCCMI is able to improve the accuracy considerably as compared to the standard algorithms. The increase in accuracy is at least minimum 15% in Spambase, 8% in CreditCard and 17% in Electricity datasets. Thus, for this test case, **Hypothesis 6 is accepted**.

### 8.2.4.2 Test for Accuracy of ASSCCMI on balanced data streams with concept drift

*Hypothesis 7:* The performance of ASSCCMI in terms of overall accuracy of balanced data streams with concept drift is better as compared to the performance of KNNADWIN, OzaBaggingADWIN and SPASC techniques on the same data streams.

**Experimental Setup:** Similar experiments are performed for balanced data streams as stated in section 8.2.4.1 above. In balanced data streams, the ratio of majority class to minority class instances is same.

**Results:**

# Research Findings and Analysis

The graphs in Figure 8-22 (a) and (b) represent the cumulative accuracy for respective percent of labeled data for each of the techniques. The results show considerable increase in the performance of ASSCCMI in terms of average accuracy as the metric.

**Hyperplane 1 Dataset**

| Percent of Labeled data in a batch | 5 | 10 | 20 | 30 | 40 | 50 | 60 | 70 | 80 | 90 |
|---|---|---|---|---|---|---|---|---|---|---|
| ASSCCMI | 0.85 | 0.84 | 0.85 | 0.84 | 0.84 | 0.84 | 0.83 | 0.85 | 0.84 | 0.85 |
| SPASC | 0.58 | 0.59 | 0.60 | 0.63 | 0.65 | 0.67 | 0.69 | 0.70 | 0.72 | 0.73 |
| KNNADWIN | 0.71 | 0.73 | 0.77 | 0.80 | 0.81 | 0.81 | 0.82 | 0.82 | 0.82 | 0.83 |
| OZABAGGINGADWIN | 0.70 | 0.72 | 0.76 | 0.77 | 0.79 | 0.79 | 0.80 | 0.80 | 0.80 | 0.81 |

(a)

**Hyperplane 2 Dataset**

| Percent of Labeled data in a batch | 5 | 10 | 20 | 30 | 40 | 50 | 60 | 70 | 80 | 90 |
|---|---|---|---|---|---|---|---|---|---|---|
| ASSCCMI | 0.89 | 0.90 | 0.89 | 0.90 | 0.90 | 0.88 | 0.89 | 0.89 | 0.89 | 0.90 |
| SPASC | 0.58 | 0.59 | 0.60 | 0.63 | 0.65 | 0.67 | 0.68 | 0.70 | 0.71 | 0.72 |
| KNNADWIN | 0.79 | 0.80 | 0.82 | 0.83 | 0.83 | 0.83 | 0.84 | 0.83 | 0.83 | 0.84 |
| OZABAGGINGADWIN | 0.77 | 0.79 | 0.81 | 0.82 | 0.82 | 0.82 | 0.82 | 0.83 | 0.83 | 0.83 |

(b)

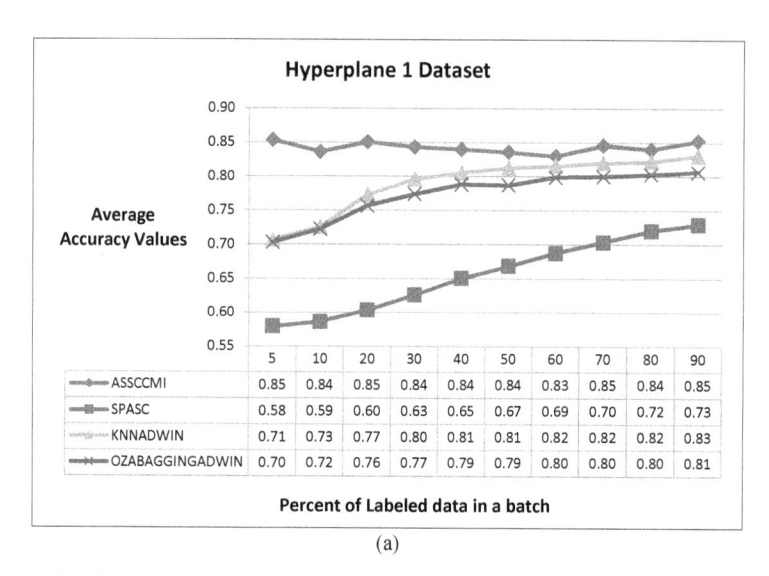

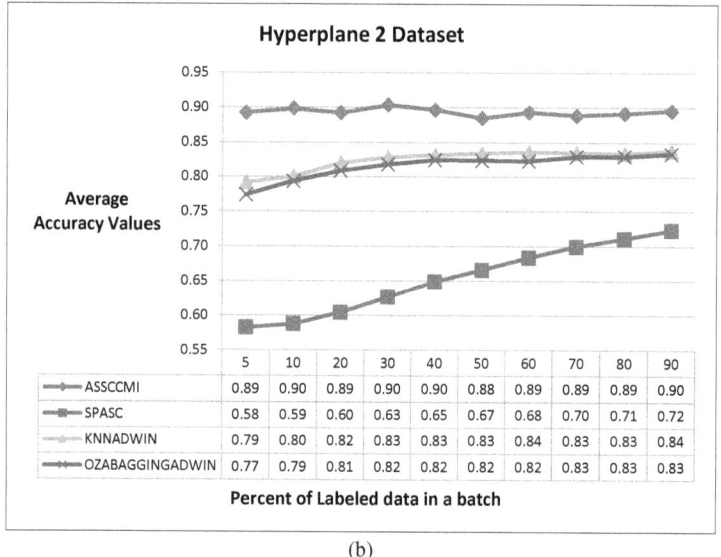

**Figure 8-22: Comparison of Average Accuracy of ASSCCMI v/s other algorithms for Hyperplane datasets**

Research Findings and Analysis

**Analysis:**

Hyperplane1 and Hyperplane2 are balanced data streams with incremental drifts which are artificially induced using MOA query (refer Appendix A.1). The accuracy of ASSCCMI for both these data streams is almost constant with a difference of more or less 1%. Since the data is balanced, equal number of micro clusters are created in a model leading to equal number in the ensemble. As a result, there is no biasing of majority class. Moreover, the drift detection mechanism is able to detect the drifts in both the data streams and maintain accuracy. The other algorithms' performance viz., of KNNADWIN and OzaBaggingADWIN, depends totally on the available labeled data. Hence for these algorithms, as the number of labeled data increases the accuracy increases. For SPASC, the performance is almost identical for both the data streams and directly proportional to the available labeled data. Since the data streams are well balanced, the percent of labeled data available affects the accuracy. Hence, it is also able to create equal clusters. Since SPASC allots max label as the label of these clusters, the more the available label the better becomes the classification.

The average accuracy of each of the algorithms for the Hyperplane datasets is shown in Table 8-11 below:

**Table 8-11: Comparison of Average Accuracy values for balanced streams with drifts**

|  | Hyperplane 1 | Hyperplane 2 |
|---|---|---|
| ASSCCMI | 84% ± 1% | 88% ± 1% |
| SPASC | 66% ± 7% | 65% ± 7% |
| KNNADWIN | 77% ± 6% | 82% ± 2% |
| OzaBaggingADWIN | 75% ± 5% | 80% ± 3% |

Thus, it can be seen that for balanced data streams ASSCCMI is able to improve the accuracy considerably as compared to the standard algorithms. The increase is accuracy is at least 7% in Hyperplane1 and 6% in Hyperplane2 datasets. Thus, for this test case also **Hypothesis 7 is accepted**.

# Research Findings and Analysis

### 8.2.4.3 Test for Accuracy of ASSCCMI on imbalanced data streams with concept drift

**Hypothesis 8:** The performance of ASSCCMI in terms of overall accuracy of imbalanced data streams with concept drift is better as compared to the performance of KNNADWIN, OzaBaggingADWIN and SPASC techniques on the same data streams.

**Experiment:** Experiments similar to the one stated in section 8.2.4.1 above are performed for imbalanced data streams with concept drifts.

**Results:**

The graphs in Figure 8-23 (a) and (b) represent the cumulative accuracy for respective percent of labeled data for each of the techniques. The results show that ASSCCMI is able to achieve considerable increase in the average accuracy for these data streams.

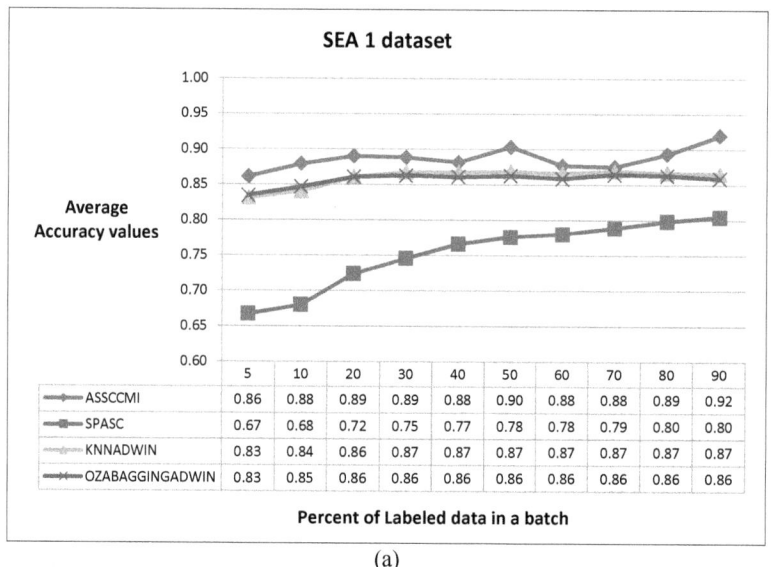

**SEA 1 dataset**

Average Accuracy values

| Percent of Labeled data in a batch | 5 | 10 | 20 | 30 | 40 | 50 | 60 | 70 | 80 | 90 |
|---|---|---|---|---|---|---|---|---|---|---|
| ASSCCMI | 0.86 | 0.88 | 0.89 | 0.89 | 0.88 | 0.90 | 0.88 | 0.88 | 0.89 | 0.92 |
| SPASC | 0.67 | 0.68 | 0.72 | 0.75 | 0.77 | 0.78 | 0.78 | 0.79 | 0.80 | 0.80 |
| KNNADWIN | 0.83 | 0.84 | 0.86 | 0.87 | 0.87 | 0.87 | 0.87 | 0.87 | 0.87 | 0.87 |
| OZABAGGINGADWIN | 0.83 | 0.85 | 0.86 | 0.86 | 0.86 | 0.86 | 0.86 | 0.86 | 0.86 | 0.86 |

(a)

Research Findings and Analysis

**SEA 2 dataset**

| Percent of Labeled data in a batch | 5 | 10 | 20 | 30 | 40 | 50 | 60 | 70 | 80 | 90 |
|---|---|---|---|---|---|---|---|---|---|---|
| ASSCCMI | 0.90 | 0.91 | 0.91 | 0.91 | 0.92 | 0.91 | 0.91 | 0.91 | 0.91 | 0.92 |
| SPASC | 0.66 | 0.68 | 0.71 | 0.73 | 0.74 | 0.75 | 0.77 | 0.78 | 0.78 | 0.78 |
| KNNADWIN | 0.84 | 0.85 | 0.87 | 0.87 | 0.88 | 0.88 | 0.88 | 0.88 | 0.88 | 0.88 |
| OZABAGGINGADWIN | 0.83 | 0.85 | 0.86 | 0.86 | 0.86 | 0.86 | 0.86 | 0.86 | 0.86 | 0.86 |

(b)

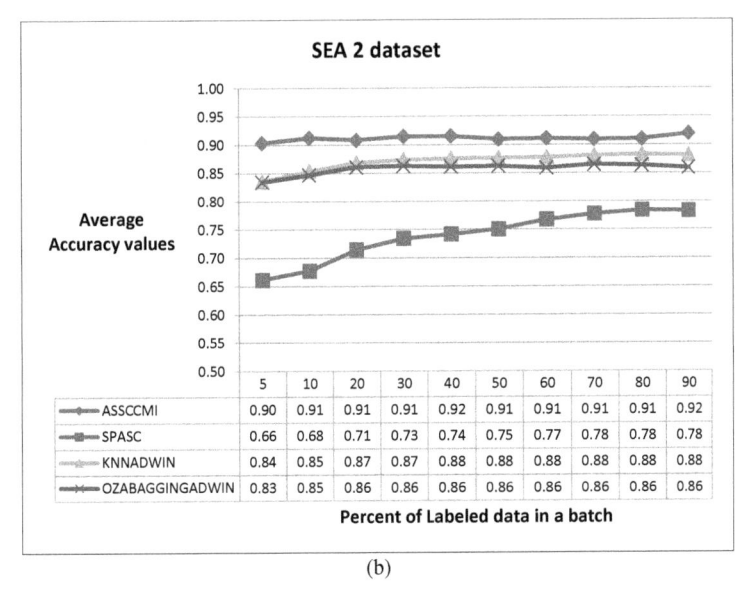

**Figure 8-23: Comparison of Average Accuracy of ASSCCMI v/s other algorithms for SEA datasets**

**Analysis:**

SEA1 and SEA2 are imbalanced data streams with imbalance ratio of 1.79 and 1.32 and with gradual and sudden drifts respectively. These streams are generated using the MOA framework (refer Appendix A.1 for MOA based data stream generation) so the exact location and type of drift is known. The accuracy measure value of ASSCCMI for SEA2 is approximately 3-4% higher than SEA1 for a particular available percent of labeled data. This is because SEA1 has gradual drift. The change in the concept happens gradually between 40000 to 60000 instances. The drift detection module is invoked only when the difference in MSE satisfies the z-test criteria. In gradual changes, the MSE of prediction is considerably low as a result the difference in accuracy between two consecutive windows is also low. The accuracy in SEA1 for ASSCCMI is better than SPASC as it only has uninformed drift detection. ASSCCMI also has better accuracy than KNNADWIN and OzaBaggingADWIN. Particularly in the case of gradual drifts, KNNADWIN seems to have performed better than OzaBaggingADWIN. In sudden drift, the z-test criterion gets satisfied most of the times as the change in the MSE is quite large. As a result, ASSCCMI can adapt better than the comparative algorithms and has better performance in SEA2. The average accuracy of each of the algorithms for the datasets is shown in Table 8-12 below:

**Table 8-12: Comparison of Average Accuracy values for imbalanced streams with drifts**

|  | SEA 1 | SEA 2 |
|---|---|---|
| ASSCCMI | 89% ± 3% | 91% ± 1% |
| SPASC | 74% ± 6% | 72% ± 6% |
| KNNADWIN | 85% ± 2% | 86% ± 2% |
| OzaBaggingADWIN | 84% ± 2% | 84% ± 2% |

Thus, it can be seen that for ASSCCMI is able to improve the accuracy considerably as compared to the standard algorithms for imbalanced data streams with drifts. The increase in accuracy is at least a minimum 4% in SEA1 and 5% in SEA2 datasets. Thus, for this test case also **Hypothesis 8 is accepted**.

### 8.2.5 Test Case for Minority class accuracy of ASSCCMI

One of the main objectives of the research is improvising learning in presence of imbalanced data streams. This test case evaluates the performance of ASSCCMI in terms of classifying the minority class in presence of imbalanced data streams and thus helps to assess the achievement of the objective. The hypothesis is defined as follows:

*Hypothesis 9:* The performance of ASSCCMI in terms of classifying the minority class in imbalanced data streams is better as compared to the performance of KNNADWIN, OzaBaggingADWIN and SPASC techniques on the same data streams.

**Experimental Setup:**

The same experimental setup as described in section 8.2.4.1 is used to evaluate this test case. The performance measure used is recall of minority class as it gives the true measure to verify the classification of minority class instances.

**Results:**

Research Findings and Analysis

The graphs shown below in Figure 8-24 to 8-27 represent the average recall values for the minority class for respective percent of labeled data for each of the techniques. The values prove that ASSCCMI is able to classify minority class instances better than the comparative algorithms.

**Analysis:**

As mentioned earlier, for each dataset, the overall imbalance ratio and the ratio in each batch of data is different. ASSCCMI is unaware of the ratio of majority to minority instances. The only information known to ASSCCMI is the class labels of majority and minority class.

**Spambase Dataset:**

Spambase Dataset — Minority class recall values

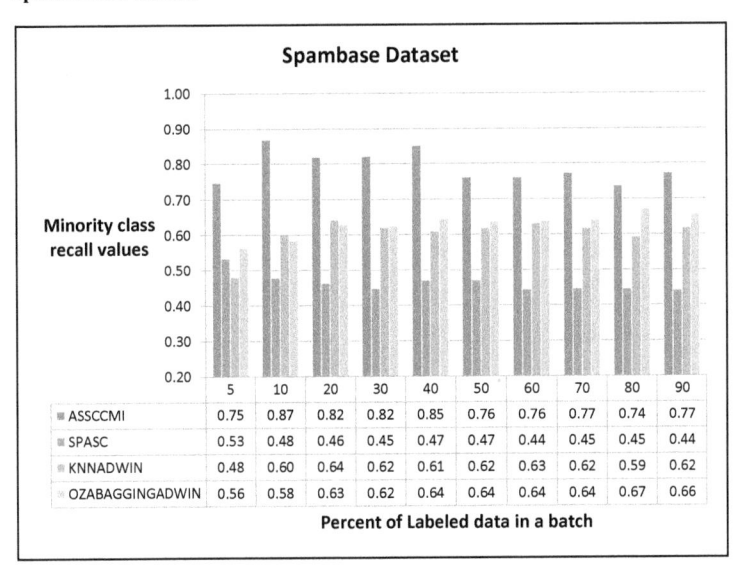

| Percent of Labeled data in a batch | 5 | 10 | 20 | 30 | 40 | 50 | 60 | 70 | 80 | 90 |
|---|---|---|---|---|---|---|---|---|---|---|
| ASSCCMI | 0.75 | 0.87 | 0.82 | 0.82 | 0.85 | 0.76 | 0.76 | 0.77 | 0.74 | 0.77 |
| SPASC | 0.53 | 0.48 | 0.46 | 0.45 | 0.47 | 0.47 | 0.44 | 0.45 | 0.45 | 0.44 |
| KNNADWIN | 0.48 | 0.60 | 0.64 | 0.62 | 0.61 | 0.62 | 0.63 | 0.62 | 0.59 | 0.62 |
| OZABAGGINGADWIN | 0.56 | 0.58 | 0.63 | 0.62 | 0.64 | 0.64 | 0.64 | 0.64 | 0.67 | 0.66 |

Figure 8-24: Comparison of Minority class recall of ASSCCMI v/s other algorithms for Spambase dataset

1. For Spambase dataset, the minority class recall is more than the other algorithms especially for lower percent of labeled data. The imbalance ratio affects the performance of all the algorithms. SPASC has the least recall values as compared to all others. KNNADWIN and OzaBaggingADWIN have a similar recall for almost all percent labeled data. The OzaBaggingADWIN gets an advantage due

# Research Findings and Analysis

to its bagging ensemble that primarily benefits the minority class. As the percent of labeled data increases the recall for OzaBaggingADWIN increases due to its online bagging technique. Due to its cluster and label approach, ASSCCMI is able to maintain minority class clusters, which is not possible for KNNADWIN and SPASC. Hence, ASSCCMI has higher recall for all of the labeled data. Moreover the accuracy is slightly high for lower percent labeled data so is the recall.

**Credit Card Dataset:**

### Credit Card Dataset

Minority class recall values

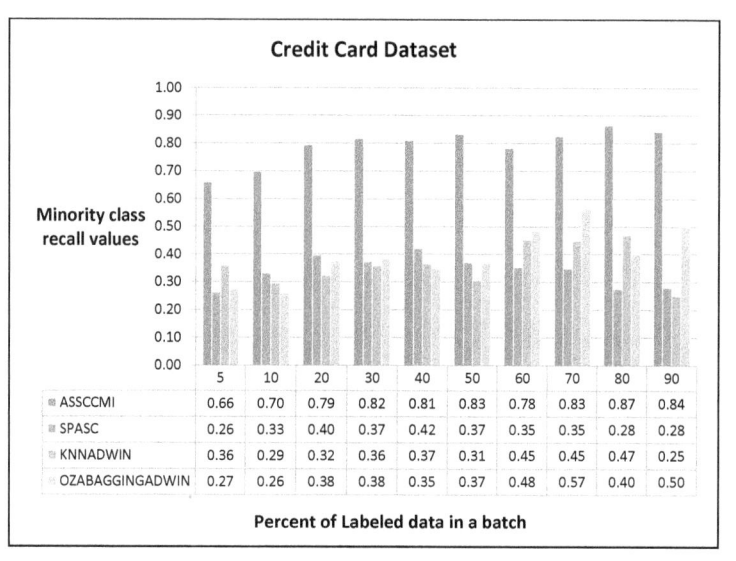

| Percent of Labeled data in a batch | 5 | 10 | 20 | 30 | 40 | 50 | 60 | 70 | 80 | 90 |
|---|---|---|---|---|---|---|---|---|---|---|
| ASSCCMI | 0.66 | 0.70 | 0.79 | 0.82 | 0.81 | 0.83 | 0.78 | 0.83 | 0.87 | 0.84 |
| SPASC | 0.26 | 0.33 | 0.40 | 0.37 | 0.42 | 0.37 | 0.35 | 0.35 | 0.28 | 0.28 |
| KNNADWIN | 0.36 | 0.29 | 0.32 | 0.36 | 0.37 | 0.31 | 0.45 | 0.45 | 0.47 | 0.25 |
| OZABAGGINGADWIN | 0.27 | 0.26 | 0.38 | 0.38 | 0.35 | 0.37 | 0.48 | 0.57 | 0.40 | 0.50 |

**Figure 8-25: Comparison of Minority class recall of ASSCCMI v/s other algorithms for Credit Card dataset**

2. For CreditCard dataset, ASSCCMI performs far better than all the other algorithms. The dataset is highly imbalanced with small data size. As a result in most of the data batches, there were very few or zero minority class instances. The classification algorithm OzaBaggingADWIN that learns from the labeled data suffers to a large extent due to this high imbalance of the data and hence has shown very poor performance. So has KNNADWIN, which depends on nearest neighbour class and due to missing minority instances in the neighbourhood it cannot classify them correctly. For SPASC, the models in its ensemble have very few representation of minority class cluster as it does not focus on the balancing

# Research Findings and Analysis

of the clusters unlike ASSCCMI. Hence it suffers majorly while classifying minority classes resulting in very poor performance. ASSCCMI, on the other hand, benefits from its unique label propagation which it performs amongst three contiguous batches. This allows ASSCCMI to leverage the availability of minority instances in any one of the three batches. As a result, it does have some micro cluster belonging to the minority class and hence is able to classify them correctly. This is evident from the increase in the minority class recall to a very high extent.

**Electricity Dataset:**

### Electricity Dataset

Minority class recall values

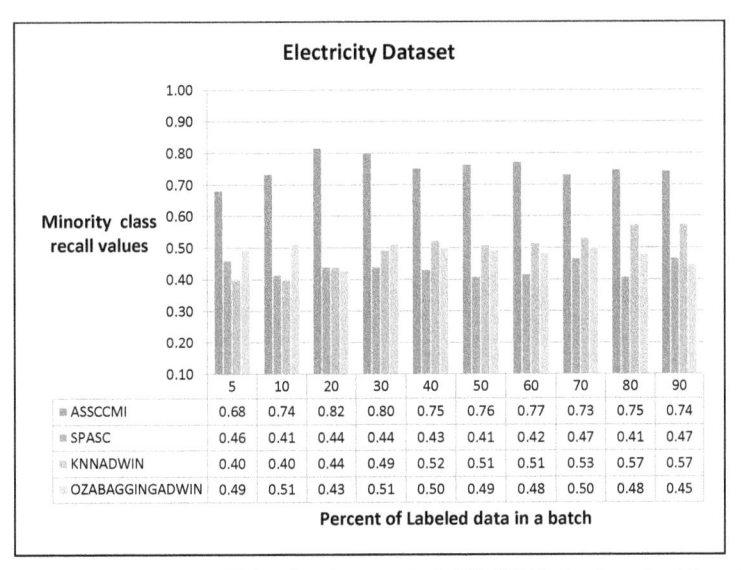

| Percent of Labeled data in a batch | 5 | 10 | 20 | 30 | 40 | 50 | 60 | 70 | 80 | 90 |
|---|---|---|---|---|---|---|---|---|---|---|
| ASSCCMI | 0.68 | 0.74 | 0.82 | 0.80 | 0.75 | 0.76 | 0.77 | 0.73 | 0.75 | 0.74 |
| SPASC | 0.46 | 0.41 | 0.44 | 0.44 | 0.43 | 0.41 | 0.42 | 0.47 | 0.41 | 0.47 |
| KNNADWIN | 0.40 | 0.40 | 0.44 | 0.49 | 0.52 | 0.51 | 0.51 | 0.53 | 0.57 | 0.57 |
| OZABAGGINGADWIN | 0.49 | 0.51 | 0.43 | 0.51 | 0.50 | 0.49 | 0.48 | 0.50 | 0.48 | 0.45 |

**Figure 8-26: Comparison of Minority class recall of ASSCCMI v/s other algorithms for Electricity dataset**

3. For Electricity dataset, ASSCCMI shows great improvement as compared to other algorithms. It was observed that, the Electricity dataset had a very high accuracy as compared to the comparative algorithms. The major reason for improved accuracy is the cluster and label approach, which is invoked only on pure clusters avoiding the majority class cluster to overlap minority class clusters. And also due to the novel cluster merge which maintains the ratio of majority and minority centroids. The clusters formed by ASSCCMI are highly cohesive and non-overlapping due to which the unseen minority class instances

are found closer to minority class clusters and the overall recall is thus higher as compared to other comparative algorithms. Moreover, the long consecutive period of majority class causes the comparative algorithms to miss out on the minority class instances. SPASC suffers a lot due to this distribution of the Electricity dataset. The implicit drifts are the third reason for improved performance of ASSCCMI. It adapts to this drift whereas the other algorithms suffer due to this. Hence, the performance of ASSCCMI is a lot better than SPASC, KNNADWIN and OzaBaggingADWIN.

**SEA Datasets:**

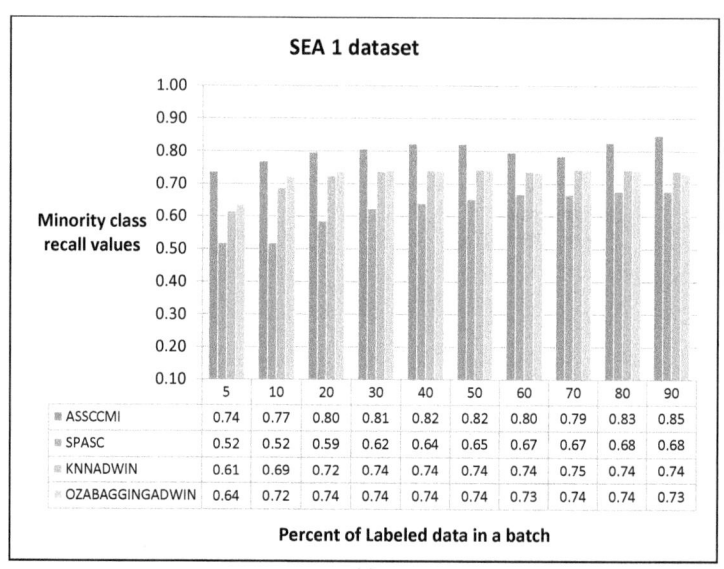

SEA 1 dataset

Minority class recall values (y-axis) vs Percent of Labeled data in a batch (x-axis)

| | 5 | 10 | 20 | 30 | 40 | 50 | 60 | 70 | 80 | 90 |
|---|---|---|---|---|---|---|---|---|---|---|
| ASSCCMI | 0.74 | 0.77 | 0.80 | 0.81 | 0.82 | 0.82 | 0.80 | 0.79 | 0.83 | 0.85 |
| SPASC | 0.52 | 0.52 | 0.59 | 0.62 | 0.64 | 0.65 | 0.67 | 0.67 | 0.68 | 0.68 |
| KNNADWIN | 0.61 | 0.69 | 0.72 | 0.74 | 0.74 | 0.74 | 0.74 | 0.75 | 0.74 | 0.74 |
| OZABAGGINGADWIN | 0.64 | 0.72 | 0.74 | 0.74 | 0.74 | 0.74 | 0.73 | 0.74 | 0.74 | 0.73 |

(a)

Research Findings and Analysis

**SEA 2 dataset**

| Percent of Labeled data in a batch | 5 | 10 | 20 | 30 | 40 | 50 | 60 | 70 | 80 | 90 |
|---|---|---|---|---|---|---|---|---|---|---|
| ASSCCMI | 0.85 | 0.86 | 0.87 | 0.87 | 0.87 | 0.88 | 0.86 | 0.87 | 0.87 | 0.88 |
| SPASC | 0.55 | 0.57 | 0.62 | 0.64 | 0.65 | 0.65 | 0.67 | 0.68 | 0.69 | 0.67 |
| KNNADWIN | 0.63 | 0.72 | 0.74 | 0.75 | 0.76 | 0.76 | 0.76 | 0.76 | 0.77 | 0.77 |
| OZABAGGINGADWIN | 0.63 | 0.73 | 0.73 | 0.76 | 0.74 | 0.76 | 0.76 | 0.77 | 0.77 | 0.77 |

(Minority class recall values)

(b)

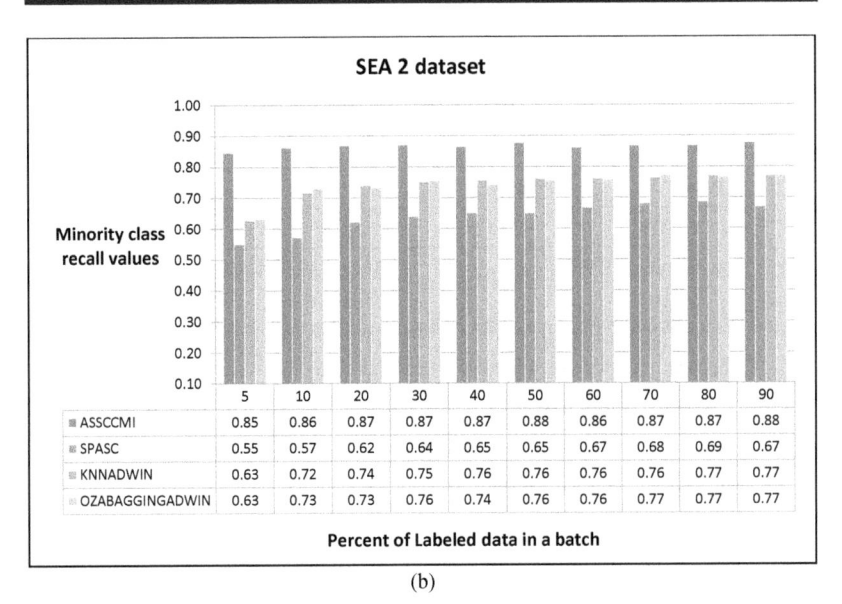

**Figure 8-27: Comparison of Minority class recall of ASSCCMI v/s other algorithms for SEA datasets**

4. For SEA1 and SEA2, the overall recall of ASSCCMI is high as compared to other algorithms. The datasets are induced with drifts which are successfully detected and adapted by ASSCCMI and is evident from the result graphs.

For SEA1 dataset; the accuracy of ASSCCMI is higher than the comparative algorithms SPASC, KNNADWIN and OzaBaggingADWIN. The recall minority class recall is quite higher depicting the ability to classify minority class better than the comparative algorithms. This is essentially due to the ability of ASSCCMI to create pure and cohesive micro clusters.

In case of SEA2 dataset, the ASSCCMI not only has higher accuracy as it is able to detect the sudden drifts in the dataset, it is also able to increase the recall of the minority class considerably as compared to the comparative algorithms. Like its constant accuracy for different percent of labeled data, ASSCCMI had constant recall of the minority class approximately around 87% which is at least 20% higher than other algorithms.

The average recall values for each of the algorithms for the datasets are shown in Table 8-13 below:

Research Findings and Analysis

**Table 8-13: Comparison of Average Minority class Recall values**

|  | Spambase | CreditCard | Electricity | SEA1 | SEA2 |
|---|---|---|---|---|---|
| ASSCCMI | 81% ± 6% | 76% ± 10% | 74% ± 6% | 80% ± 5% | 87% ± 1% |
| SPASC | 49% ± 4% | 34% ± 8% | 44% ± 3% | 60% ± 8% | 62% ± 7% |
| KNNADWIN | 57% ± 8% | 38% ± 9% | 49% ± 9% | 68% ± 6% | 70% ± 7% |
| OzaBaggingADWIN | 61% ± 6% | 38% ± 12% | 47% ± 4% | 69% ± 5% | 70% ± 7% |

Thus, it can be seen that for imbalanced data streams ASSCCMI is able to improve the recall values considerably as compared to the standard algorithms. The increase is approx. minimum 20% in Spambase, 37% in CreditCard, 25% in Electricity datasets, 11% in SEA1 and 17% in SEA2. **Thus, for this test case too, the Hypothesis 9 is accepted.**

### 8.2.6 Test Cases for Execution Time of ASSCCMI

*Hypothesis 10:* ASSCCMI requires same or less time of execution as compared to KNNADWIN, OzaBaggingADWIN and SPASC.

**Experiment:**

The average time taken to classify for each of the algorithms is calculated using in built functions from Python. For ASSCCMI and SPASC, the model is trained first which is referred to as training time. For ASSCCMI, the time required to process and train the model for first 6 batches is considered as the training time whereas for SPASC the time required to create the first model is considered as its training time. For KNNADWIN and OzaBaggingADWIN the labeled data from the first 6 batches are used for training, which is considered as their training time. The classification time for all algorithms is calculated as the time taken to predict class labels.

**Results:**

The time required for training the model and then to classify a batch is shown in Table 8-14 and Table 8-15 respectively. The lower range in the time specifies the time required for training 5% labeled and the higher range specifies 90% labeled data. The ranges in classification time specify the minimum and maximum time required for classifying a batch.

# Research Findings and Analysis

**Table 8-14: Comparison of Training time by ASSCCMI v/s other algorithms**

| Dataset Name / Batch size | ASSCCMI Training time (6 * batch) instances | SPASC Training time (1st batch) instances | KNNADWIN Training time (6 * batch_size) instances | OzaBagging ADWIN Training time (6 * batch_size) instances |
|---|---|---|---|---|
| Credit Card / 100 instances | 7.7 – 9.4s | 4.8 – 11.5s | 0.2 – 3.0s | 0.04 – 0.38s |
| Spambase / 100 instances | 12 – 14s | 10 – 24s | 0.06 – 1.25s | 0.11 – 2.61s |
| Hyperplane1 / 1000 instances | 70 – 102s | 9 – 37s | 0.4 – 14.3s | 0.3 – 3.2s |
| Hyperplane2 / 1000 instances | 76 – 109s | 9 – 57s | 0.4 – 16.4s | 0.4 – 2.9s |
| SEA1 / 1000 instances | 69 – 86s | 4.3 – 279s | 0.2 – 8.17s | 0.2 – 2.0s |
| SEA2 / 1000 instances | 69 – 97s | 10 – 344s | 0.3 – 9.13s | 0.3 – 2.1s |
| Electricity / 1000 instances | 102 – 117s | 16 – 406s | 0.2 – 3.1s | 0.2 – 1.9s |

**Table 8-15: Comparison of Classification time by ASSCCMI v/s other algorithms**

| Dataset Name / Batch size | ASSCCMI | SPASC | KNNADWIN | OzaBagging ADWIN |
|---|---|---|---|---|
| Credit Card / 100 | 0.11 – 0.2s | 4.9 – 7s | 0.15 – 0.2s | 0.02 – 0.2s |
| Spambase / 100 | 0.09 – 0.3s | 10 – 19s | 0.01 – 0.2s | 0.02 – 0.8s |
| Hyperplane1 / 1000 | 0.9 – 1.1s | 12 – 18s | 0.06 – 2.6s | 0.07 – 0.8s |
| Hyperplane2 / 1000 | 0.9 – 1.5s | 10 – 21s | 0.06 – 2.5s | 0.09 – 0.6s |
| SEA1 / 1000 | 0.8 – 1.5s | 5 – 90s | 0.04 – 1.4s | 0.06 – 0.4s |
| SEA2 / 1000 | 0.9 – 1.8s | 4.8 – 84s | 0.07 – 1.9s | 0.13 – 0.3s |
| Electricity / 1000 | 0.3 – 3.7s | 16 – 37s | 0.05 – 0.5s | 0.07 – 0.6s |

**Analysis:**

The average time taken to classify a batch when ASSCCMI is trained is far less than SPASC. ASSCCMI first builds the cluster based classifier from the first 6 batches, then refines, and updates the model while classifying. However, SPASC starts classifying as soon as a single model is trained and updates model along with classification. Hence the classification time of SPASC is considerably more than ASSCCMI.

ASSCCMI however requires more time than KNNADWIN and OzaBaggingADWIN. This is because these algorithms employ prequential evaluation and the training of the model involves use of only labeled data. So for low percent labeled data the time required is very less. As the percent of labeled data increases, the time to train also increases.

ASSCCMI requires very less time as compared to SPASC but takes more time as compared to KNNADWIN and OzaBaggingADWIN. ASSCCMI takes more time for training the model and retraining the model when drift is detected, so wherever a drift is detected, the model takes extra time for retraining and refining the ensemble. The classification time required by ASSCCMI is almost constant. Also for all the algorithms, as the batch size increases the training and classification time will also increase proportionately.

**8.3   SUMMARY**

This chapter discusses about the various test cases and hypothesis that were designed to meet the research objective. Mapping of research objectives with test cases helped to identify research objectives that were fully/partially met. The models were tested on a range of datasets which were either real or artificially constructed to justify problem statement. The results discussed above prove that the proposed model ASSCCMI outperforms the comparison models in terms of overall accuracy and recall. The results also prove that the model is able to detect and adapt to sudden, gradual and incremental drifts. The model ASSCCMI also outperforms semi-supervised algorithm SPASC in terms of execution time. The percent improvement and the analyses are discussed in detail for each topic. It can be positively concluded that ASSCCMI achieves its research objectives successfully and solves the problem it stated for different data sets.

CPSIA information can be obtained
at www.ICGtesting.com
Printed in the USA
BVHW051510170423
662378BV00028B/808